AAPS Introductions in the Pharmaceutical Sciences

The *AAPS Introductions in the Pharmaceutical Sciences* book series is designed to support pharmaceutical scientists at the point of knowledge transition. Springer and the American Association of Pharmaceutical Scientists (AAPS) have partnered again to produce a second series that juxtaposes the *AAPS Advances in the Pharmaceutical Sciences* series. Whether shifting between positions, business models, research project objectives, or at a crossroad in professional development, scientists need to retool to meet the needs of the new scientific challenges ahead of them. These educational pivot points require the learner to develop new vocabulary in order to effectively communicate across disciplines, appreciate historical evolution within the knowledge area with the aim of appreciating the current limitations and potential for growth, learn new skills and evaluation metrics so that project planning and subsequent evolution are evidence-based, as well as to simply "dust the rust off" content learned in previous educational or employment settings, or utilized during former scientific explorations. The *Introductions* book series will meet these needs and serve as a quick and easy-to-digest resource for contemporary science.

Ajay Babu Pazhayattil • Sanjay Sharma
Joe Paul Philip • Michelle Gischewski-Silva
Marzena Ingram

Technology Transfer

Drug Product Manufacturing Process

Springer

aaps

Ajay Babu Pazhayattil
cGMPWorld
Welland, ON, Canada

Sanjay Sharma
Zydus Group
Ahmedabad, Gujarat, India

Joe Paul Philip
Glenmark Pharmaceuticals
Mumbai, Maharashtra, India

Michelle Gischewski-Silva
Eurofins CDMO
Mississauga, ON, Canada

Marzena Ingram
Validant Inc.
San Francisco, CA, USA

ISSN 2522-834X ISSN 2522-8358 (electronic)
AAPS Introductions in the Pharmaceutical Sciences
ISBN 978-3-031-32191-7 ISBN 978-3-031-32192-4 (eBook)
https://doi.org/10.1007/978-3-031-32192-4

This Springer imprint is published by the registered company Springer Nature Switzerland AG
The registered company address is: Gewerbestrasse 11, 6330 Cham, Switzerland

Forewords

This book *Technology Transfer: Drug Product Manufacturing Process* is a valuable resource for individuals seeking to gain practical knowledge and insightful guidance on pharmaceutical technology transfer. It delves into the complexities of this process and presents readers with strategies to overcome challenges. The book emphasizes the opportunities that technology transfer presents, such as improving manufacturing efficiency and reducing costs, accelerating product development timelines, and expanding capabilities. It offers guidance on leveraging opportunities by implementing best practices in project planning, risk management, and stakeholder engagement. It is an essential guide for professionals in the pharmaceutical industry who want to navigate technology transfer successfully.

Prashant Sharma, Chief Technical Officer at Zydus Life Sciences
Ahmedabad, India

As I begin to reflect on the importance of pharmaceutical technology transfer, I am struck by the vital role it plays in the development and delivery of safe and effective medications. In the ever-evolving world of pharmaceuticals, it is crucial to have a resource that provides comprehensive information on this critical process. This book is exactly that resource. It is a comprehensive guide for anyone involved in pharmaceutical technology transfer, from researchers and developers to manufacturers and regulators. The authors have brought together a wealth of knowledge and expertise, providing a detailed overview of the entire process, from initial research and development to full-scale production.

Dr. Rajiv Desai, Indian Pharmaceutical Alliance
Mumbai, India

The success of technology transfer in the pharmaceutical industry depends on a deep understanding of the science and technology involved. This includes knowledge of the drug, as well as the processes used to manufacture them. In addition, it requires expertise in regulatory compliance, quality, and project management. The book provides a comprehensive and authoritative guide to the key principles and practices of technology transfer in the pharmaceutical industry. It provides practical advice and best practices that can be applied in real-world situations. The book is an indispensable resource for anyone working in technology transfer.

Vipul Doshi, Head of Quality at Zydus Life Sciences
Ahmedabad, India

Preface

Technology transfer in the pharmaceutical industry is a critical process that ensures the safe and efficient transfer of knowledge, skills, and technology from one organization to another. This process is essential for the development and commercialization of drug products and for the scaling up of existing products to meet the needs of patients worldwide. The process of technology transfer involves the integration of a wide range of activities, including process development, equipment qualification, scale-up, process performance qualification, continued process verification, as well as the training of personnel. It also includes the transfer of knowledge from R&D to manufacturing and from one facility to another. The importance of technology transfer in the pharmaceutical industry cannot be overstated, as it plays a crucial role in ensuring the quality, safety, and efficacy of drugs for patients.

The successful execution of technology transfer requires a thorough understanding of the process, as well as the ability to navigate the many regulatory requirements, quality control measures, and technical considerations. The process also requires the coordination and collaboration of multiple teams, departments, and even companies. To minimize the risk of errors, delays, or non-compliance, it is crucial to have a detailed plan, clear communication, and good documentation practice. Therefore, technology transfer can be tricky, and it is crucial to have experienced professionals who are familiar with the process and understand the technical and regulatory requirements.

This book aims to provide a comprehensive overview of technology transfer in the pharmaceutical industry and to provide readers with a detailed understanding of the process, including the key considerations and best practices. It is intended for students, pharmaceutical industry aspirants, and professionals working in the pharmaceutical industry, including technical operations, manufacturing science & technology, validation, R&D, manufacturing, quality assurance, quality control, and regulatory affairs. The book aims to be a comprehensive guide to technology transfer, providing readers with the knowledge and tools necessary to navigate the complexities of the process successfully and to ensure that the quality, safety, and efficacy of drugs for patients are maintained throughout.

As authors, we have put our best efforts into creating a comprehensive and informative guide on technology transfer. We have drawn on our extensive experience and knowledge in the field to provide readers with a detailed understanding. We have included the technical points that must be considered during the technology transfer process. There are some real-world examples to help readers better understand the concepts presented. We believe that this book will be an invaluable resource for professionals. We thank Dr. Claudio J. Salomon, our well-wishers, and mentors, for wholeheartedly supporting this initiative. Special acknowledgment to Itaan Pharma and Zydus Life Sciences for providing the figures used. We are confident that the information and guidance provided in this book will help readers steer technology transfer and ensure an efficient transfer. We have put an effort into making the book easy-to-understand as possible, and we hope you will enjoy reading it.

cGMPWorld
Welland, ON, Canada
Ajay Babu Pazhayattil

Zydus Group
Ahmedabad, Gujarat, India
Sanjay Sharma

Glenmark Pharmaceuticals
Mumbai, Maharashtra, India
Joe Paul Philip

Eurofins CDMO
Mississauga, ON, Canada
Michelle Gischewski-Silva

Validant Inc.
San Francisco, CA, USA
Marzena Ingram

Contents

About the Authors

Dr. Ajay Babu Pazhayattil is an experienced pharmaceutical professional with a strong background in industrial pharmacy and executive management. He has a track record of driving innovation for brand, generic, biosimilar, and CDMO small molecule/biologic organizations. He is well-respected in the industry for his data-driven, regulatory-compliant QbD based strategies. Dr. Pazhayattil is an expert in injectable and solid oral manufacturing process- design, characterization, technology transfer, qualification, and continued process verification. He is actively involved in industry organizations and is recognized as a thought leader in the field, having made significant contributions through published journal articles and textbooks.
https://www.linkedin.com/in/pazhayattil/

Sanjay Sharma is a seasoned pharmaceutical industry professional with over 23 years of experience in the development, launch, and maintenance of drug supply. He possesses a results-driven approach and a deep understanding of pharmaceutical manufacturing science, which has enabled him to contribute to the success of some of the largest organizations in the industry. Sanjay is the Senior Vice President & Head of Manufacturing Science and Technology, Zydus Group. He has held various leadership positions in Indian and multinational pharmaceutical companies such as Cipla, Dr. Reddy's Laboratories, Sandoz, Watson, Torrent, and Lupin.
https://www.linkedin.com/in/sanjay-sharma-5787109/

Joe Paul Philip has over 20 years of experience in technical and strategic leadership roles, specifically in life cycle management of formulations. He has experience leading multicultural teams globally and expertise in process development, new product launches, technical services, R&D technology transfers, global technology transfers, process validations, computer system validations, equipment qualifications, technical investigation, root cause analysis, and data analytics for OSD, parenteral, and semi-solid technology platforms. Joe Paul is the Vice President &

Head of Global Process Excellence and New Products, Glenmark Pharmaceuticals. He also has experience and knowledge in negotiating development/commercial agreements.
https://www.linkedin.com/in/joe-paul-philip-5478629/

Michelle Gischewski-Silva As a highly experienced and accomplished leader in the pharmaceutical industry, Michelle brings a wealth of knowledge and expertise to her role. With a solid background in pharmaceutical science, and expertise with critical new product development projects, she possesses deep technical knowledge in R&D formulation development, process development, technology transfer, scale-up, and qualification of solid and liquid dosage forms. As the head of the formulation, she is responsible for optimizing novel formulations and streamlining processes to ensure robustness.
https://www.linkedin.com/in/michelle-gischewski-74734915/

Marzena Ingram is an experienced independent senior pharmaceutical consultant who specializes in quality and technical operations for solid-dose and active pharmaceutical ingredient manufacturing. She helps clients address FDA warning letter scenarios. Marzena has developed specialized teams for implementing continued process verification. She has also implemented programs to meet global regulatory requirements, developed and published statistical tools for pharmaceutical manufacturing processes, and led the implementation of a comprehensive life cycle management software. Marzena is the current Vice President of ISPE Canada Executive Board.
https://www.linkedin.com/in/marzena-ingram-a387b09b/

Chapter 1
Current Status of Technology Transfer

Abstract The textbook is a practical guide for technology transfer, MS&T, and technical operations departments in organizations. The book prioritizes examples of technology transfers from R&D to own sites, as this is where most transfers occur. Chap. 1 discusses the key findings of a survey conducted as part of completing the textbook. The survey acknowledges that Knowledge Management (KM) tools are well established in the pharmaceutical industry. Based on the data, the need for a special section on post-approval change management protocol (PACMP) was determined. The industry has not fully adopted ICH Q12 concepts. The textbook emphasizes the importance of defining deliverables from the sending unit (SU) to the receiving unit (RU) to ensure successful technology transfer. This book provides tools to help identify the necessary information for successful transfers. It also uses well-understood definitions and does not delve into less standard terms (as determined by the survey) like KMA and KPP. The authors encourage industry-wide harmonization of methods for measuring successful technology transfer projects. The survey results indicate that the industry is increasingly using software tools for a data-driven approach.

Keywords Technology Transfer · Manufacturing Science and Technology · Technical Operations · Knowledge Management · Success · Data Driven

Introduction

An industry survey on Pharmaceutical Drug Product Technology Transfer (Stage 1b) was executed by the authors to gauge the current practices in the industry. Since the existence of ICH Q8, organizations have had various levels of adoption phases. The survey results, in fact, provide insights into the practical application levels of quality by design (QbD) [1] and lifecycle concepts during technology transfer, post a maturity period of almost 15 years since its existence. The industry survey focused on the key stakeholders, i.e., innovators, generic/biosimilar organizations, contract development and manufacturing organizations (CDMO), and contract research organizations (CRO). The survey instrument and authors ensured that the identity, organization, department, and demography of the participants will not be published.

A. B. Pazhayattil et al., *Technology Transfer*, AAPS Introductions in the Pharmaceutical Sciences 10, https://doi.org/10.1007/978-3-031-32192-4_1

The participant information was only utilized to validate and assure that the data collected were from seasoned technology transfer industry professionals.

Survey Results

Though the survey was not designed to capture certain percentages of each group, the participants in fact reflected the technology transfer/manufacturing volumes. In total, 71% of the participants were from generic/biosimilar organizations and 17% were from CDMO organizations (Fig. 1.1).

Based on the participant responses, it was identified that 73% of the technology transfer activities are undertaken by the Technology Transfer/MS&T group of the organization (Fig. 1.2). Only 10% identified the R&D group as responsible for technology transfer and scale-up activities (Stage 1b) in their organization. The

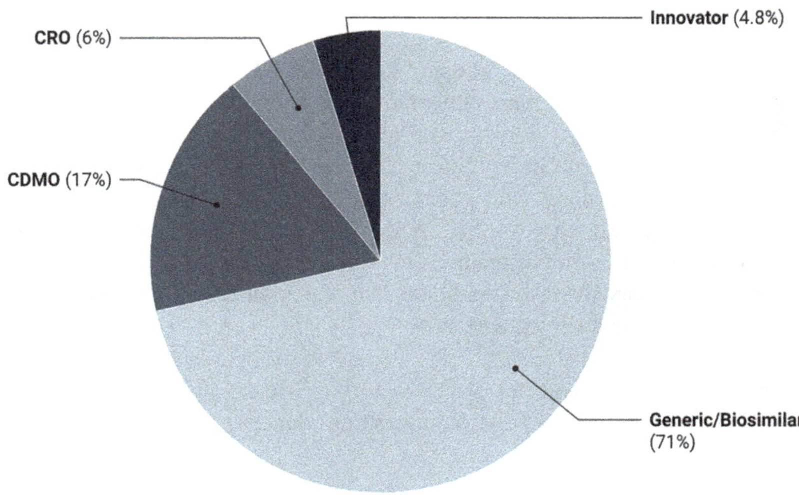

Fig. 1.1 Technology transfer survey participant: industry segments

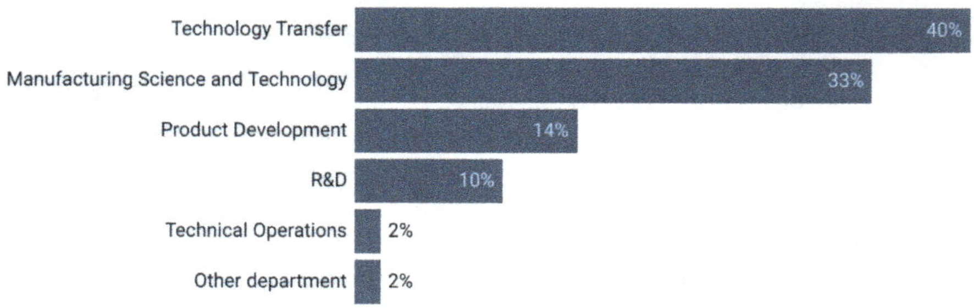

Fig. 1.2 Department responsible for technology transfer and scale-up

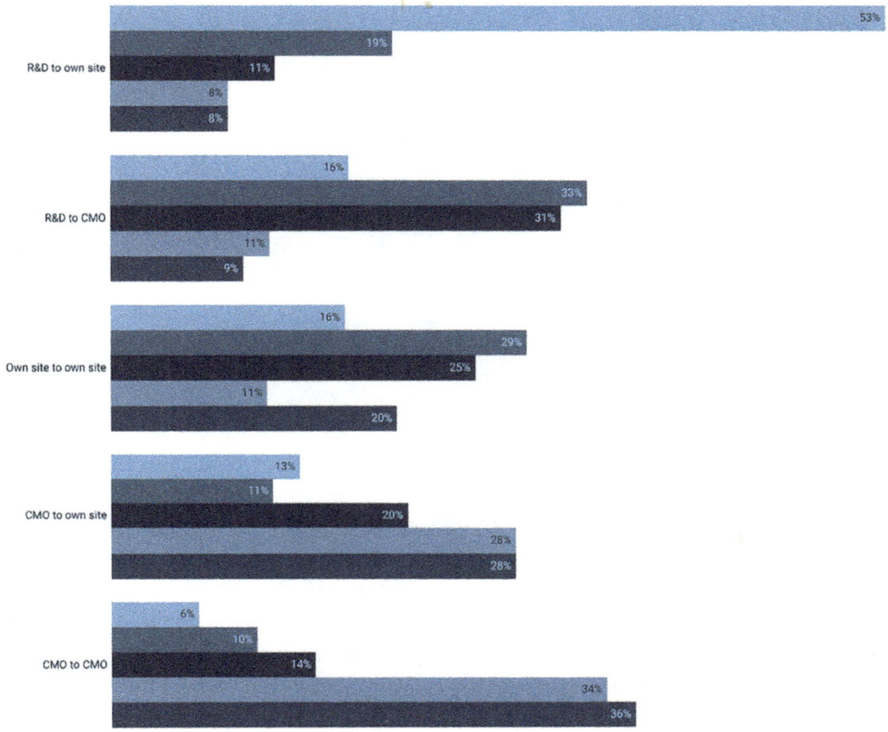

Fig. 1.3 Sending unit and receiving unit

response clearly indicates that most organizations now have a dedicated department (within or independent of R&D/Product Development) for technology transfer and scale-up.

Technology transfer can happen between an R&D site and a contract manufacturing organization (CMO) or the organization's own manufacturing site. Often organizations transfer between their own manufacturing locations. There are scenarios where the process is transferred between two CMOs as well as from a CMO to the organization's own manufacturing site. These scenarios were posed and ranked to decipher the most occurring transfers. R&D to own site was ranked as the most occurring scenario, while CMO to CMO transfer was the least (Fig. 1.3). The data aligns with the fact that most of the generic/biosimilar organizations undertaking a high volume of technology transfer projects typically transfer the manufacturing technology from their own R&D site to their own manufacturing site.

The US FDA process validation guidelines [2] encourage using previous credible experience with sufficiently similar products and processes. The survey, therefore, gauged the current utilization of such prior data to understand the effects of the scale and establish operating ranges. All the respondents (100%) identified using previous credible data from sufficiently similar products and processes during their technology transfer projects (Fig. 1.4). Of these, 33% apply the data for more than 90% of

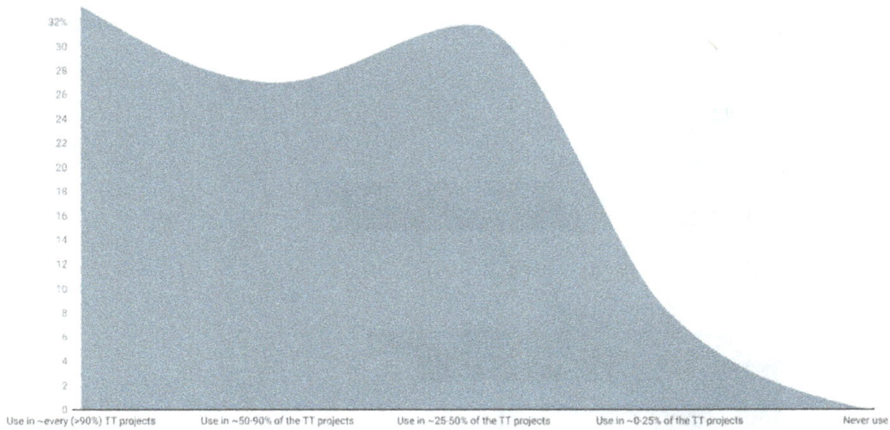

Fig. 1.4 Use of previous credible data

their technology transfer projects. Only 8% noted that they utilize past data for less than 25% of their projects. The responses reflect the maturity of Knowledge Management (KM) tools [3] in the pharmaceutical industry.

The relatively recent ICH Q12 Guideline [4] complements the existing ICH Q8, Q9, Q10, and Q11 [5–7] Guidelines and provides a framework to facilitate the management of post-approval Chemistry, Manufacturing, and Controls (CMC) changes. The tools associated with the guidelines provide organizations to make post-approval changes in a more predictable and efficient manner across the product lifecycle. The swift adoption of ICH Q12 is expected to benefit the end patient and pharmaceutical organizations. The survey gauged the adoption of post approval change management protocol (PACMP) or comparability protocol to allow flexible continuous improvement changes per ICH Q12. The survey result shows that the adoption of ICH Q12 concepts is inconsistent, where only less than 50% apply for their new product launches. The data indicates that organizations may need to mandate the adoption for their own benefit (Fig. 1.5). Fourteen percent do not consider utilizing the pay-off benefit offered by the regulator for products developed utilizing the QbD approach.

Technology transfer involves multiple activities and departmental interactions. The challenges can be from the testing lab to manufacturing operations related or product related as well. The survey, therefore, asked the participants to identify and rank the most challenging factor out of the most common issues during a technology transfer project. The common issues identified were scheduling the operations, availability of resources, test method transfer/validation, developing design of experiments (DoE's), and knowledge transfer gaps. The survey identified the knowledge transfer gap as the most challenging issue during technology transfer (Fig. 1.6). Developing and executing the DoE studies itself was the least challenging. The survey result points toward the need for establishing the deliverables from sending unit (SU) to receiving unit (RU) and the constant involvement of the two parties involved.

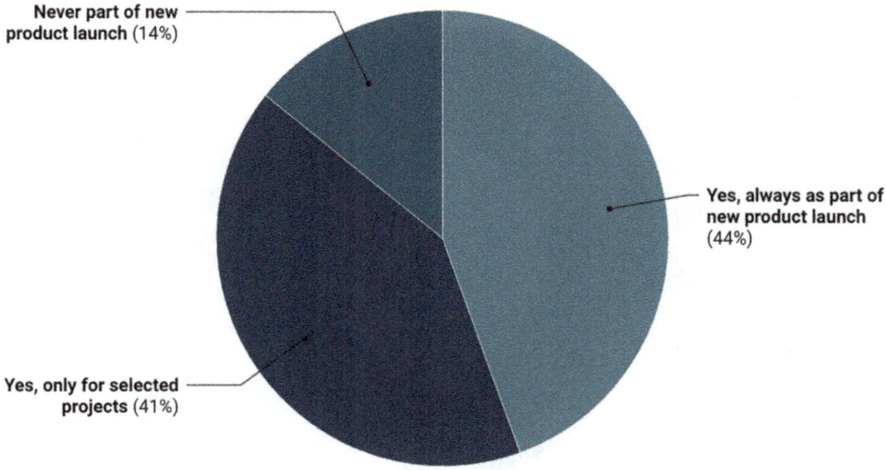

Fig. 1.5 ICH Q12 adoption

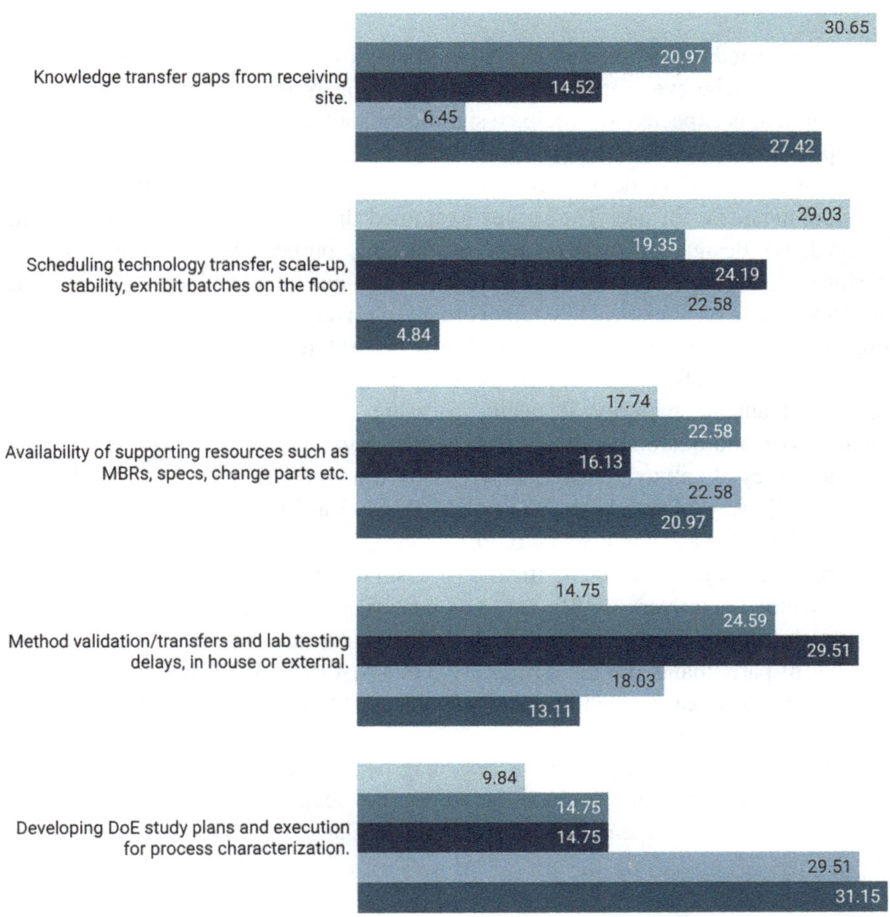

Fig. 1.6 Technology transfer challenges

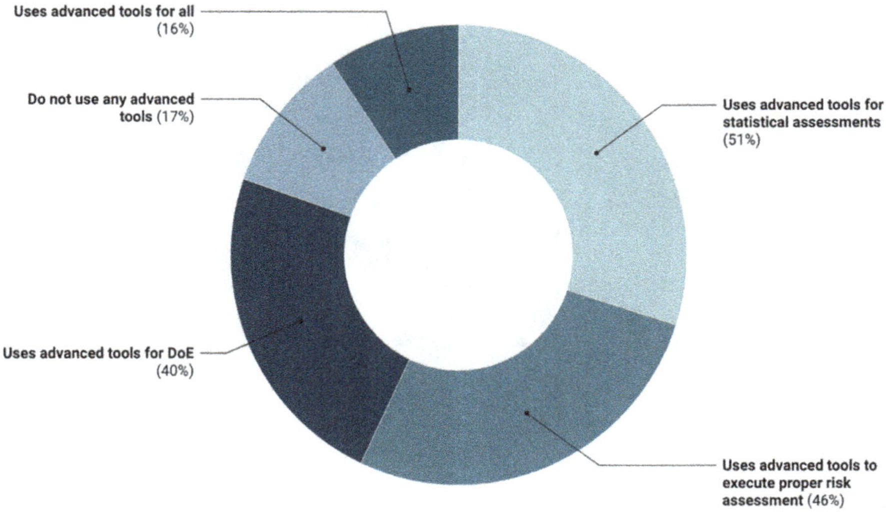

Fig. 1.7 Use of advanced tools

It is expected for pharmaceutical firms to employ objective measures and statistical metrics wherever feasible and meaningful at Stage 1. Process design and development is expected to anticipate significant sources of variability and establish the appropriate detection, control, and mitigation strategies. Therefore, the use of quantitative, statistical methods is imperative. Many statistical tools and techniques are now available for use. The survey examined the use of advanced tools for the rationalized design of experiments, performing proper risk assessments, and performing multivariate statistical analysis. Fifty-one percent identified as using advanced tools for statistical assessment (e.g., multivariate tools). Fourty-six percent use advanced risk assessment tools. Forty percent used tools for DoE studies. However, only 16% use advanced tools in all three areas (Fig. 1.7). The results reflect advancing maturity in using software tools for a quantitative, semi-quantitative, and qualitative data-driven approach to technology transfer.

The survey attempted to estimate the use of the terms: Key Performance Parameters (KPP), Key Material Attributes (KMA), Critical Material Attributes (CMA), and Critical Process Parameters (CPP). Through guidance, referring to Critical Material Attributes (CMA), Critical Process Parameters (CPP), the terms Key Performance Parameters (KPP), and Key Material Attributes (KMA) were understood to be inconsistently used within the industry. As anticipated, an overwhelming number of participants use the terms CPP and CMA (98% and 84%). Only 31% of the survey participants noted that they identify KMA for materials in their technology transfer projects (Fig. 1.8).

A technology transfer project success may be measured by the product/process robustness measures put in place. However, the reading can be taken after the completion of Stage 1 and before moving into Stage 2, after the completion of

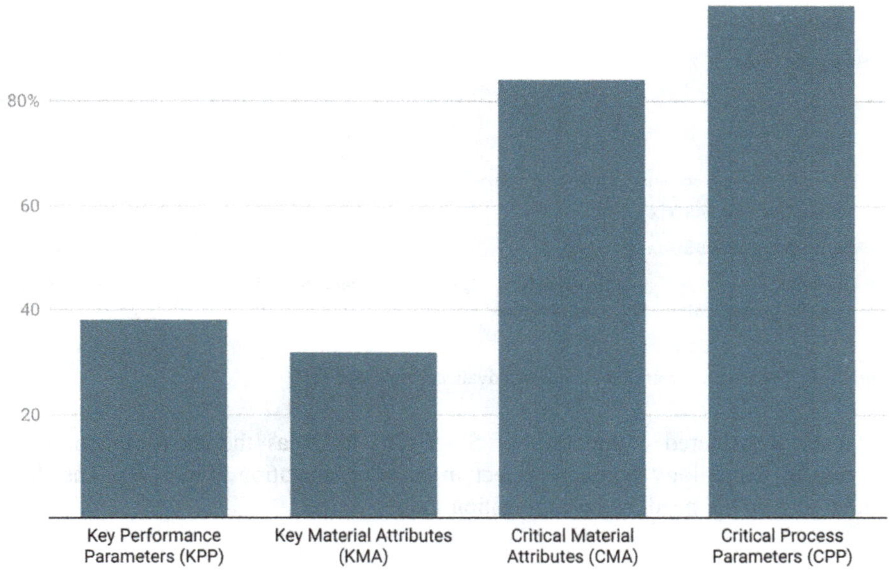

Fig. 1.8 Terms CPP, CMA, KPP, KMA

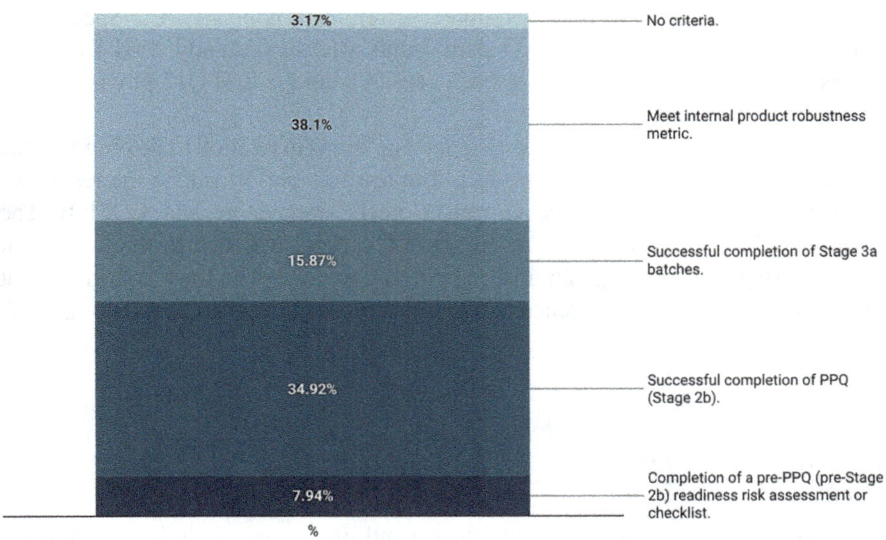

Fig. 1.9 Measuring the success of technology transfer

Stage 2b (PPQ), or after the completion of Stage 3a. Organizations may have their own criteria or not have any criteria at all. Based on an examination of the survey participant responses, it is now understood that most organizations have their own internal gauge of the success of a technology transfer project. Thirty-five percent,

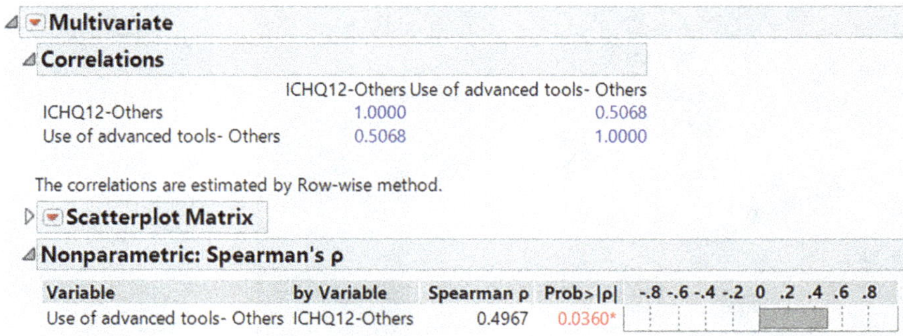

Fig. 1.10 Spearman's correlation: use of advanced tools and Q12

however, considered completion of Stage 2b (PPQ) as the measurement of a successful technology transfer project in their organization (Fig. 1.9). The data points toward the need for harmonization industrywide.

Additional Insights

A positive correlation was observed upon performing a nonparametric Spearman's correlation. The innovator, CDMO participants who consistently used DoE, risk assessment, and statistical tools, seemed to be utilizing the ICH Q12 provisions to a larger extent (Fig. 1.10).

A t-test was conducted on the participant's opinions in relation to determining the success of a technology transfer project. The test was performed for the responses from generic industry participants versus others (innovators and CDMO). The responders had varied opinions, as in, the TT success was tied to the success of stage 2b, stage 3b, per stage 2b, internal criteria, or none. The t-test confirmed that there was no significant difference in opinion across innovators, CDMOs or generic organizations (Fig. 1.11).

Conclusion

The subsequent chapters will attempt to expand on the gaps identified during the survey and target the areas that will give the most benefit to the readers. The chapters focused on generic/biosimilar process technology transfer considering the volumes. Since most organizations now have a dedicated department for technology transfer, the subsequent chapters will only focus on the activities conducted by the Technology Transfer/MS&T department. Since the greatest number of technology transfers are happening between R&D to own site, such examples will be prioritized. It is now

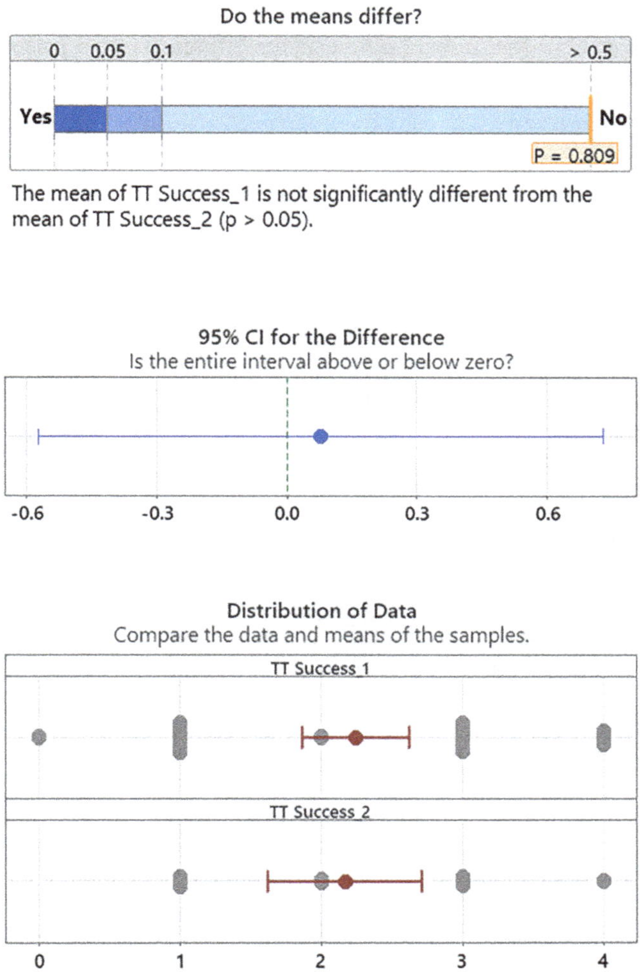

Fig. 1.11 Measuring success of a TT project

understood that Knowledge Management (KM) tools exist in the pharmaceutical industry, and hence less effort will be given to explain the common tools used. The adoption of ICH Q12 concepts is not at the fullest; therefore, a special section on utilizing post-approval change management protocol (PACMP) is included. Establishing the deliverables from sending unit (SU) to receiving unit (RU) has been identified as the key to a successful technology transfer. The succeeding chapters will identify checklists and tools to identify the types of information enabling successful transfers. The terms KMA and KPP are less common, and hence the textbook will continue to stick with the well-understood definitions only. The survey results indicate that the industry is maturing in using software tools for a quantitative, semi-quantitative, and qualitative data-driven approach to technology transfer. The authors also encourage industrywide harmonization of methods to measure a successful technology transfer project.

References

1. ICH (2009) Q8(R2) pharmaceutical development. In: International conference on harmonization of technical requirements for registration of Pharmaceuticals for Human use considerations
2. FDA (2011) Process validation: general principles and practices. Center for Drug Evaluation and Research (CDER)
3. ISPE (2021) Good practice guide: knowledge Management in the Pharmaceutical Industry. International Society for Pharmaceutical Engineering
4. ICH (2019) Q12 technical and regulatory considerations for pharmaceutical product lifecycle management. In: International conference on harmonization of technical requirements for registration of Pharmaceuticals for Human use considerations
5. ICH (2023) Q9(R1) quality risk management. In: International conference on harmonization of technical requirements for registration of Pharmaceuticals for Human use considerations
6. ICH (2008) Q10 pharmaceutical quality system. In: International conference on harmonization of technical requirements for registration of Pharmaceuticals for Human use considerations
7. ICH (2012) Q11 development and manufacture of drug substances (chemical entities and biotechnological/biological entities). In: International conference on harmonization of technical requirements for registration of Pharmaceuticals for Human use considerations

Chapter 2
Quality by Design (QbD) Process Design

Abstract Quality by design (QbD) principles have become widely adopted in the pharmaceutical industry to ensure consistent quality of products through their formulation, process development, and commercialization. Advanced analytical techniques, including spectroscopy, chromatography, microscopy, and thermal analysis, are vital in controlling the critical product and process parameters necessary to produce high-quality pharmaceutical products consistently. Manufacturers also employ advanced technologies such as continuous manufacturing to improve efficiency and lower costs. Quality risk management (QRM) principles are critical in identifying, evaluating, and controlling risks throughout the development process, including research, development, and manufacturing. Computer-aided process simulation has become more prevalent in optimizing pharmaceutical manufacturing processes. Understanding the science behind the product and the manufacturing process is crucial, as knowledge gained from early formulation and process development stages, including scale-up, is integral. Overall, pharmaceutical formulation and process development require a multidisciplinary approach incorporating advanced analytical techniques, manufacturing technologies, and QbD and QRM principles.

Keywords Quality by Design (QbD) · Quality Risk Management (QRM) · Characterization · Design of Experiments (DoE)

Introduction

It is essential to assess the ability of a manufacturing process to endure the anticipated or unanticipated variability of input raw materials (API and excipient), processing conditions (equipment and environment), and human factors to produce with desired preset specifications consistently.

> The lifecycle concept links product and process development, qualification of the commercial manufacturing process and maintenance of the process in a state of control during routine commercial production. – USFDA Guidance on Process Validation 2011.

A. B. Pazhayattil et al., *Technology Transfer*, AAPS Introductions in the Pharmaceutical Sciences 10, https://doi.org/10.1007/978-3-031-32192-4_2

The guidance indicates a strong emphasis on a lifecycle approach instead of an event-based one. Though there may be differences in terminologies used across various guidance, the overarching principle of lifecycle management and the use of knowledge gained as part of the product development journey to be the basis for process improvement and innovation forms the central theme of guidance. The objective of an effective process design is to ensure that the drug product is produced as fit for its intended use, adhering to the preestablished target product profile.

A Quality-by-Design Approach

Quality by design (QbD) is a systematic approach to pharmaceutical development that begins with predefined objectives and emphasizes product and process understanding and control based on sound science and quality risk management. The goal of QbD is to ensure that the quality of the final product is built into the design of the manufacturing process and that the process is capable of consistently delivering quality products. QbD (Fig. 2.1) can be applied throughout the development and manufacturing process, from the early stages of product development to the commercial manufacturing of the product. Developing a robust product with desired quality attributes that are consistently met requires a systematic approach to product development.

QbD involves a systematic and scientific approach to pharmaceutical development that begins with predefined objectives and emphasizes product and process understanding and control based on sound science and quality risk management. It involves designing and controlling the manufacturing process to produce a product

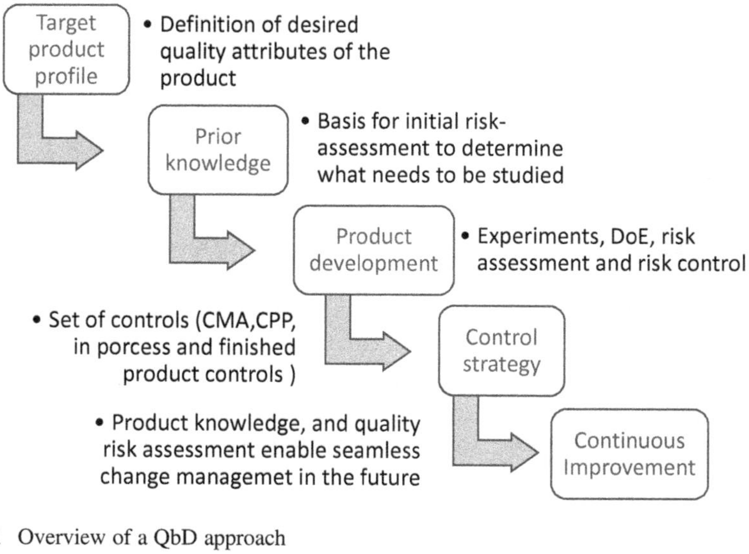

Fig. 2.1 Overview of a QbD approach

that meets its predetermined quality attributes consistently. The QbD process includes the following steps:

- Define the product and process design: Identify the quality target product profile (QTPP) and critical quality attributes (CQAs) of the product, as well as the critical process parameters (CPPs) that have an impact on the CQAs.
- Develop and design the process: Use the understanding of the product and process to design the manufacturing process.
- Verify the process design: Conduct appropriate testing to ensure that the process is capable of consistently producing a product that meets its predetermined quality attributes.
- Control the process: Establish control strategies and monitoring systems to ensure that the process remains in a state of control and consistently produces a product of the desired quality.
- Continual improvement: Use a systematic approach to identify and evaluate opportunities for process improvement and make necessary changes to enhance process performance and product quality.

QbD has been shown to improve product quality, reduce manufacturing variability and defects, and increase process efficiency and predictability. It is increasingly being adopted as a regulatory requirement for the development and manufacture of pharmaceutical products.

Active Pharmaceutical Ingredient (API), the Excipients, and the Packaging Components

To understand how the physical and chemical properties of the active pharmaceutical ingredient (API) may impact the performance of the final drug product, it may be necessary to conduct studies on the drug product itself. These studies can provide valuable information that can be used to set appropriate specifications for the API and to ensure that it is compatible with any excipients in the final product. If the drug product contains multiple APIs, it is also essential to evaluate the compatibility of these ingredients. This information can help to ensure that the drug product is safe and effective for use in patients.

Polymorphism, the ability of a chemical compound to exist in more than one form or crystal structure, can have significant effects on the physical and chemical properties of the compound, including its solubility, stability, and bioavailability. Polymorphism can be an essential consideration in drug formulation, as the properties of the API can significantly influence the performance of the final product. If the API has low solubility, it may be difficult to achieve sufficient drug levels in the body, affecting the drug's effectiveness. Therefore, identifying and selecting a specific form of the API most suitable for the intended use is critical. Characterization methods (Fig. 2.2), such as PXRD and DSC [1, 2]., are performed to explore if changes occur in the active ingredient solid-state characteristics.

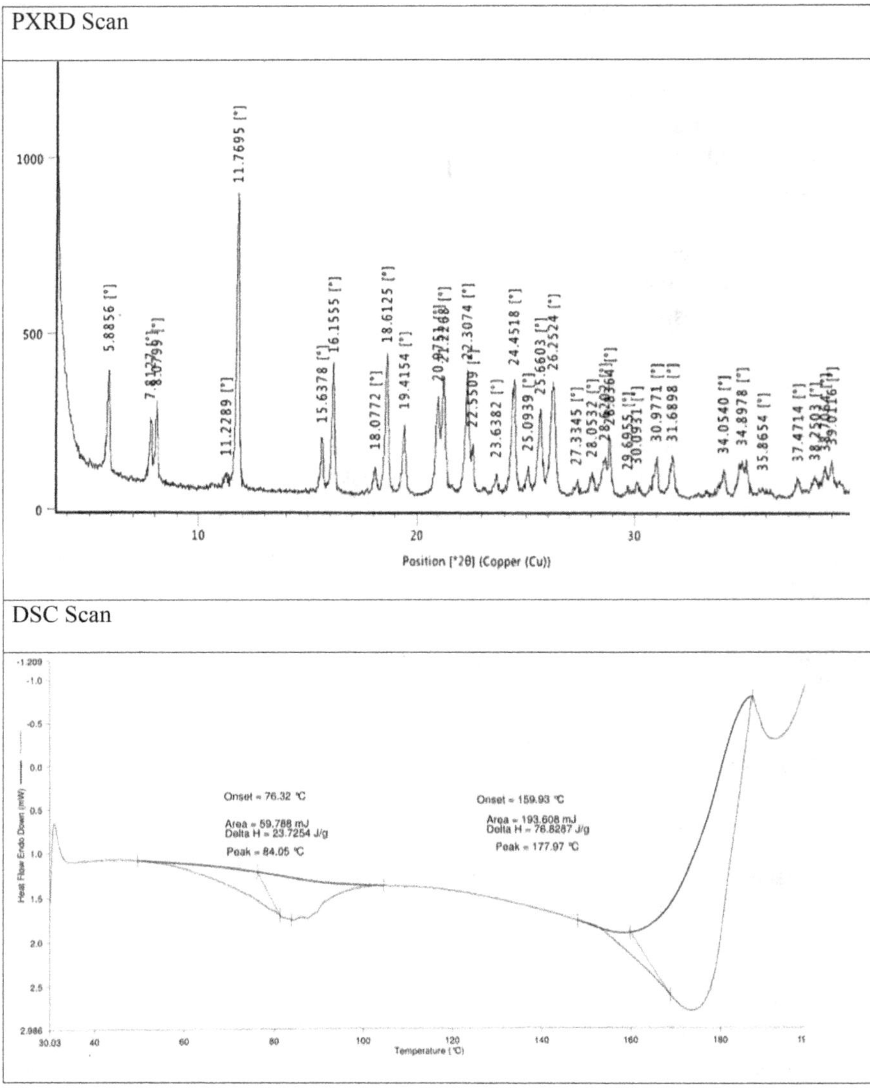

Fig. 2.2 Examples of PXRD, DSC characterization peaks

Drug substance properties can be classified as physical, chemical, and biological. A list of the most relevant properties of the drug substance that are to be considered, along with their potential impact, is provided in Table 2.1.

It is essential to carefully select and evaluate the excipients used in the drug product, considering their concentration and any characteristics that may affect its performance or manufacturability. This includes all substances used in the manufacturing process, even if they do not appear in the final product (e.g., processing aids). It is also essential to assess the compatibility of excipients with

Table 2.1 Example of drug substance properties assessed

Physical properties	**Impact**
Salt and polymorph	It can affect solubility and stability
Melting point	Determines consistency at room temperature and potential for change during processing
Solubility—Aqueous solubility across the pH range	Essential to understanding its dissolution state and availability for absorption in different segments of the gastrointestinal tract
Appearance—Particle shape, particle size, and distribution	Can affect the solubility rate, process performance
Density	Critical to processability at different unit operations
Flow properties	Critical for processability, mainly when the drug load in the dosage form is high. If processability is poor, it must be altered through the manufacturing process
Hygroscopicity	Determines precautions during the manufacturing process and defines the packaging configuration
Chemical properties	**Impact**
Chemical stability in solid state	Crystalline and amorphous have different stability. It will define if significant stability issues (degradation) should be expected during the product shelf life. Appropriate measures to suppress the degradation would be selected accordingly
Chemical stability in solution	Drug substance in solution represents worst-case scenario for chemical reactivity and aids in understanding the degradation pathways as a function of pH. Understanding the specifics of pH sensitivity will aid in designing a stable drug product by applying a control over the pH of the microenvironment in the drug product that is favorable to maintaining a stable drug substance. Excipients are carefully chosen to aid the chemical stability of API and the final formulation
Oxidative stability	Oxidative stress on the drug substance will reveal the potential for oxidative degradation. Selection of the ingredients will be narrowed to those which do not have oxidative properties. Protection from oxygen from the air may be needed during the manufacturing process or in packaging. Alternatively, adding an antioxidant to stabilize the product toward oxidation may be needed
Photosensitivity	Sensitivity to light is important to understand the potential contribution of exposure to light on degradation
pKa	To understand the dissociation of the functional groups at different pH and the impact on solubility
Biological properties	**Impact**
Partition coefficient	Defines affinity of the drug substance toward hydrophilic or lipophilic media. Determined as a *log* ratio of the amount found in octanol vs. water, it can indicate the absorption ability of the active, which is driven by the balance between the hydrophilicity and lipophilicity of the compound
Membrane permeability	Indicates absorption affinity in vivo. Could be determined in vitro (e.g., Caco-2 cell layer) or in vivo in animals (intestinal perfusion). Human PK studies, where mass balance or absolute bioavailability are determined, can be used as an indication of GIT permeability

(continued)

Table 2.1 (continued)

Biopharmaceutics Classification System (BCS) class	Solubility and permeability in terms of BCS: BCS 1. High Solubility and High Permeability BCS 2. Low Solubility and High Permeability BCS 3. High Solubility and Low Permeability BCS 4. Low Solubility and Low Permeability Solubility is classified based on single-dose solubility in 250 ml of media across the pH range (high <250 ml, low >250 ml). Low-soluble drugs are more challenging as the dissolution rate is critical to achieving bioavailability. The dissolution rate can be significantly impacted by formulation. High permeability is defined based on the extent of absorption (more than 85–90% of the administered dose)
Pharmacokinetic properties	They represent the overall impact of drug substance physical, chemical, and biological properties in in vivo environment. Drug substance properties heavily drive absorption, distribution, metabolism, and elimination

each other and the active pharmaceutical ingredient as relevant. The ability of excipients to provide their intended functionality and to maintain that functionality throughout the shelf life of the product should also be demonstrated. The information obtained from these evaluations can be used to support the selection and quality of the excipients and to justify the specifications for the final drug product. The Inactive Ingredient Database contains details about the inactive ingredients found in drugs approved by the FDA [3]. This information can be helpful in the process of creating new drug products. If an inactive ingredient has already been used in an approved drug via a defined dosage delivery, it is tried and tested and may not need as much scrutiny the next time it is used in a new drug.

It is essential to use a data-driven, scientific approach when developing and qualifying a packaging solution for a pharmaceutical product. This includes engaging with stakeholders to identify all the product-related quality requirements and using risk management and quality by design (QbD) principles (Fig. 2.3) to develop and select a suitable solution. The solution should consider factors such as the supplier's reliability and technical capabilities, as well as the product's logistical parameters, such as temperature, transit routes, and storage conditions. The process should also include design testing, technical reviews, and traceability. By following these steps, it is possible to develop a robust acceptance criterion for the packaging solution that takes into account the specific needs of the product.

Formulation Development

Formulation development outlines the process of creating a formulation that meets desired product performance criteria and can be consistently produced on a commercial scale. This process starts with a formulation risk assessment (Table 2.2). The first step in the formulation development process is a risk assessment to identify factors that could impact the critical quality attributes (CQAs) (Fig. 2.4). This

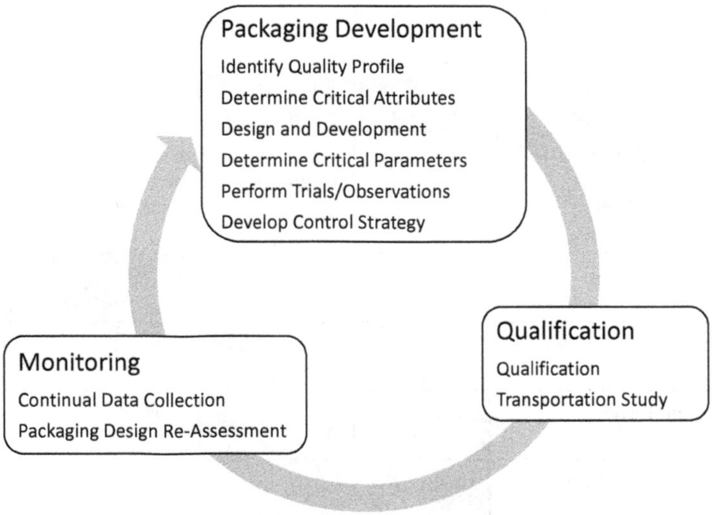

Fig. 2.3 QbD packaging lifecycle

assessment can utilize a cause-and-effect technique and draws on previous knowledge of the drug substance, the process, for example, immediate-release tablets, and relevant unit operations such as wet granulation, drying, blending, and compression. The results of this assessment are then used to assign quantitative risk priority numbers to potential issues, which are categorized as high, medium, or low risk.

A risk evaluation is then conducted on the chosen excipients. In addition to API-excipient compatibility studies [4] for the key excipients in the formulation, risks such as the ones listed can be identified (Table 2.3).

Several different formulation approaches are typically considered during the development process. The approach that is chosen depends on the solubility of the active ingredient. For a drug substance with low solubility, possible approaches may include the following:

- Organic solvent granulation, where the active is dissolved.
- Aqueous/hydroalcoholic hydro-organic granulation approach, where active is in powder bed.
- Hot melt granulation with a melting ingredient.
- Spray drying to convert the crystalline form into an amorphous form.
- Dry granulation with micronized drug substance.

Formulation Studies

Small-scale trials are executed with various approaches. Hot melt granulation can cause impurity issues and may not improve the dissolution sufficiently. Dry granulation may not result in desired dissolution profile if the particle size is not the only controlling mechanism for the release rate. The organic and aqueous-based

Table 2.2 An initial formulation risk assessment

Formulation Composition Material Attributes	Drug Product CQAs			
	Assay	Impurities	Content Uniformity	Dissolution
Active Particle Size	Low	Low	High	High
Granulating Solvent Type	Low	High	Low	Low
Disintegrant Level	Low	Low	Low	Medium
Solubilizer Level/Type	Low	Low	Low	High
Binder Level	Low	Low	Low	High
Diluent Level/Grade	Low	Low	Medium	Medium
Lubricant Level	Low	Low	Low	Medium
Glidant Level	Low	Low	Low	Medium

Relative Risk Ranking:

Low Risk	*No further investigation is needed*
Medium Risk	*Further investigation may be needed*
High Risk	*Further investigation is needed*

granulations may offer additional advantages, such as solubilizing the low-soluble drug substance.

Description of a typical formulation development trial: The reduced particle size of active and presence of surface-active agent was appropriate for a formulation of a low soluble drug. Various organic solvents was considered based on the solubility and stability of the drug substance over the period and under different temperatures that would correspond to the use of that solvent in the granulation process. Based on the drug substance stability in the solution, an optimal solvent was selected. Then, the formulation trials were designed and conducted based on this approach. However, the dissolution profile of the trial composition, under discriminatory conditions

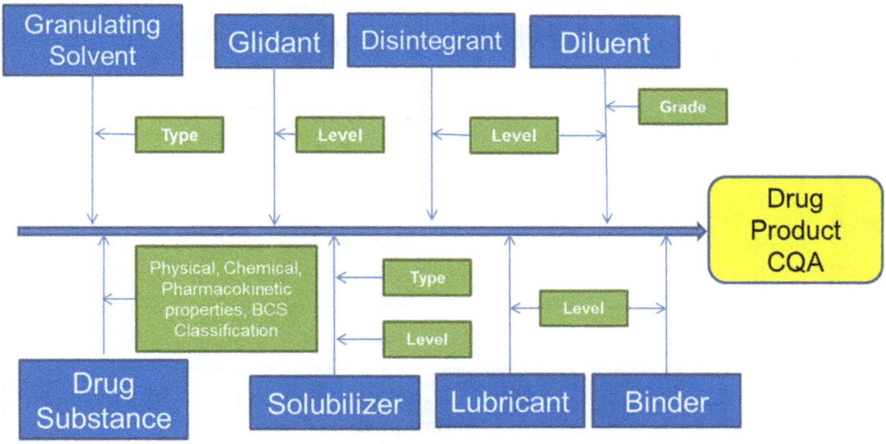

Fig. 2.4 Formulation variables that can affect the CQA

Table 2.3 Risk assessment justification

Formulation attributes	Justification
API particle size	Critical to the dissolution profile and content uniformity
Granulating solvent type	It can potentially impact the impurity profile
Disintegrant level	It can potentially impact disintegration and dissolution
Solubilizer level/type	It can potentially impact the dissolution profile
Binder level	It can potentially impact disintegration and dissolution
Diluent level/grade	It can potentially impact processability and blend homogeneity. It can impact disintegration and dissolution
Lubricant level	It can potentially impact processability and dissolution
Glidant level	It can potentially impact processability and dissolution

selected for the early development stage, were not satisfactory, i.e., dissolution is still lower than the reference product. This was a concern for achieving bioequivalence. The reduction of particle size of the drug substance by dissolving it in an organic solvent, along with the presence of surfactant in the composition, was found not sufficient to achieve a similar drug release to the innovator's product. It was hypothesized that the solubilization process is deemed critical. Hence, another approach whereby composition contains solubilizing agent and is manufactured using an aqueous (hydro-organic) based granulation process was evaluated in further development trials.

Consequently, the aqueous-based granulation approach was selected for further development trials.

To decide on the appropriate API particle size for the aqueous granulation approach, trials using active of different particle sizes are conducted and subjected to dissolution (Fig. 2.5). Based on the outcome, the particle size of the API will be

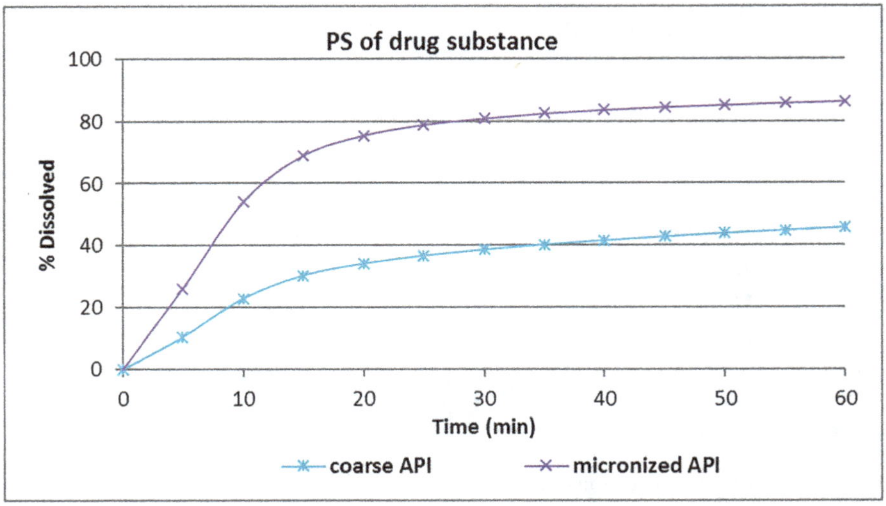

Fig. 2.5 Effect on the particle size of drug substance on dissolution

defined. If dissolution is incomplete, clear particle size reduction (e.g., micronization) is required.

Other formulation studies include:

Effect of the binder: The effect of the binder in the composition is optimized by conducting trials with different levels and measuring the effect on the tablet's physical properties and dissolution.

Effect of solubilizer: The solubilizer type used in the innovator product was adopted. The level of the selected solubilizer is optimized through the evaluation of the dissolution as an attribute most responsive to the level change.

Effect of disintegrant: The type of disintegrant is selected based on the product.

Effect of diluent: Diluents, in general, are less likely to have an impact on dissolution; however, they can impact processability. Nevertheless, when the amount in the formulation is high and the drug substance is poorly soluble, selecting the type is critical for both processability and dissolution. The type is to be selected considering the compatibility studies. The grade is also studied to select the more suitable one for the poorly soluble drug substance and the selected aqueous-based granulation manufacturing process. Based on processability and dissolution, the final grade selection is made.

Based upon the formulation studies executed at a small scale, composition and manufacturing processes are selected. This formulation is to be taken to the process evaluation stage, whereby the trials are executed at an intermediate scale, to see if reproducible results can be obtained. At this initial process evaluation stage, some critical process parameters can be preliminarily evaluated. For example, the formation of degradation products, which may be related to the duration of the process, may be measured as a function of processing time. This is important to know before commencing the full process optimization. At this stage, additional changes to the process and/or composition are typically made to improve processability or product performance. Once satisfactory performance is obtained at each processing stage and on the finished product, additional dissolution testing is conducted on selected compositions in various pH to evaluate dissolution similarity to the reference product across the pH range to confirm the suitability of the formulation and process. Based on the formulation development studies, the composition and process are tentatively finalized. They are confirmed to be able to produce tablets with desirable, target quality attributes, and as such, the product is ready to move to the process optimization stage. Acceptable ranges for the high-risk attributes are established and included in the control strategy. Based on the results of the formulation development studies, the formulation risk assessment is updated.

Process Development

Process parameters are identified as critical when a change can result in failure to meet the QTPP. Process parameters are not critical when there is no trend to failure and no evidence of significant interactions within the proven acceptable range (PAR). Trials are carried out to establish appropriate control strategies to minimize the effects of variability in material attributes and process parameters on CQAs.

Design of Experiments (DoE)

To propose ranges for above-mentioned critical process parameters, a DoE [5] is applied. General knowledge and experience with dosage forms, manufacturing processes, drug substance, and excipient characteristics are used only as a starting point in designing experiments for a specific product under development. This set of experiments aids in gaining knowledge and understanding of the particular product, both material and process-related attributes. At the experimental stage, selected critical elements of the materials (the drug substance and proposed excipients) along with manufacturing process parameters at each processing step are varied systematically in predefined design to confirm its impact on the product performance, i.e., on product critical quality attributes (quality target product profile—QTTP). Ultimately, the study's outcome proposes a proper control strategy over critical material attributes (CMA) and critical process parameters (CPP) that will ensure the quality profile of the drug product is consistently met, and there is no significant variability.

As an example, for wet granulation mixing, DoE study impeller/chopper speeds, flow rate, and kneading time are known to be critical process parameters of the finished dosage unit. Utilization of 2^{3-1} fractional factorial DoE to evaluate the

Table 2.4 Batch flow for 2^{3-1} fractional factorial design

Factors		Coded values		
		−1	0	+1
		Actual values		
A: Spray rate (g/min)		(Low)	(Target)	(High)
B: Impeller/chopper speed (rpm)		(Low)/(Low)	(Target)/ (Target)	(High)/(High)
C: Kneading time		(Low)	(Target)	(High)
WTG	Parameters for granulation			
	Premix	Liquid addition		Wet mixing
	Impeller/chopper speeds (rpm)	Impeller/chopper speeds (rpm)	Flow rate (g/min)	Kneading time (min)
Trial 1	(Low)/Off	(Low)/(Low)	(High)	(Low)
Trial 2	(Low)/Off	(Low)/(low)	(Low)	(High)
Trial 3	(High)/Off	(High)/(High)	(High)	(High)
Trial 4	(High)/Off	(High)/(High)	(Low)	(Low)
Trial 5	(Target)/Off	(Target)/(Target)	(Target)	(Target)

influence of critical process parameters, i.e., impeller/chopper speed, solution flow rate, and kneading time, on the physical and analytical performance parameters of finished dosage units and their effect on desired product profile is applied (Table 2.4).

The physical and chemical attributes of granules are loss of drying (LOD)—in process, residual solvent (organic volatile impurity—OVI), water content, degradation products, granule particle size distribution, bulk and tapped density, and blend uniformity. As wet granulation involves evaporation of the hydro-organic solvent after the granulation step, samples are collected throughout the drying stage to monitor the level of residual solvents and thus evaluate the efficacy of the drying process. The OVI and water contents from all optimization batches are evaluated in connection to the processing parameters (impeller/chopper speeds, spray rate, and kneading time). As heat is used for drying, degradation products are also monitored throughout the drying process to evaluate the effect of heat on degradation. Based on impurity results, OVI and % water, an optimal drying time and in-process LOD limit are finalized. Considering the in-process LOD results, OVI and RC results, appropriate controls for the drying step, are recommended. Following the completion of the drying process, the dried material is milled through the selected screen and then blended for particle size uniformity. Samples are removed after blending and evaluated for sieve profile and bulk density. Physical properties (particle size and density) and blend uniformity results from all the DoE batches are assessed.

A statistical model incorporating interactive terms can be used to evaluate the effect of the independent variables on the dependent variables. An example of the impact of granulation parameters on granule attributes is shown in Fig. 2.6.

As shown in the above charts, processing parameters impeller/chopper speed, flow rate, and kneading time within the studied range do not significantly impact

% retained on Screen #1 (coarse) Sorted Parameter Estimates

Term	Estimate	Std Error	t Ratio	t Ratio	Prob>|t|
Impeller/Chopper Speed	-0.065	0.064846	-1.00		0.4992
Spray Rate	0.035	0.064846	0.54		0.6849
Kneading Time	0.025	0.064846	0.39		0.7657

% retained on Screen #2 (intermediate) Sorted Parameter Estimates

Term	Estimate	Std Error	t Ratio	t Ratio	Prob>|t|
Spray Rate	-1.83	1.815687	-1.01		0.4975
Impeller/Chopper Speed	1.27	1.815687	0.70		0.6114
Kneading Time	-0.49	1.815687	-0.27		0.8322

% below Screen #2 (fine) Sorted Parameter Estimates

Term	Estimate	Std Error	t Ratio	t Ratio	Prob>|t|
Spray Rate	2.055	1.889477	1.09		0.4733
Impeller/Chopper Speed	-1.425	1.889477	-0.75		0.5886
Kneading Time	0.505	1.889477	0.27		0.8337

Bulk Density Sorted Parameter Estimates

Term	Estimate	Std Error	t Ratio	t Ratio	Prob>|t|
Impeller/Chopper Speed	0.0365	0.015205	2.40		0.2513
Kneading Time	0.0175	0.015205	1.15		0.4554
Spray Rate	0.008	0.015205	0.53		0.6917

% Water Sorted Parameter Estimates

Term	Estimate	Std Error	t Ratio	t Ratio	Prob>|t|
Spray Rate	-0.05	0.290689	-0.17		0.8916
Impeller/Chopper Speed	0	0.290689	0.00		1.0000
Kneading Time	0	0.290689	0.00		1.0000

Residual Solvent Sorted Parameter Estimates

Term	Estimate	Std Error	t Ratio	t Ratio	Prob>|t|
Impeller/Chopper Speed	-12.25	7.490828	-1.64		0.3494
Kneading Time	-3.75	7.490828	-0.50		0.7045
Spray Rate	0.25	7.490828	0.03		0.9788

Fig. 2.6 Granulation parameters

granulation properties (particle size, bulk density, % water, and residual solvent). Since performing wet granulation at different settings can affect the quality attributes of the granulation, this can also affect the quality attributes of the finished dosage form. After defining the parameter ranges that ensured the granulation could be successfully manufactured, granulations from the DoE study are used to compress into tablets and evaluated against proposed acceptance criteria obtained during optimization of the compression stage. Particularly, granulations manufactured at extremes of wet granulation settings can result in tablets that display differences in physical attributes such as hardness and appearance, and analytical attributes such as dissolution can be affected (Fig. 2.7).

As shown in the above charts, granulation processing parameters impeller/chopper speed, flow rate, and kneading time within the studied range do not significantly impact the tableting properties (hardness, thickness, and friability) as well as analytical results (assay and dissolution). Granulation parameters of impeller/chopper speed, spray rate, and kneading time within the studied ranges resulted in granules with similar physical and chemical properties. The physical and chemical properties of the blend and finished dosage form produced from the granules were within the acceptance criteria. Consequently, based on the acceptable physical parameters and critical product attributes of in-process and finished dosage form, the operating ranges for granulation parameters are proposed for submission (exhibit) batches with high sheer granulation.

Analytical Quality by Design (AQbD) Based Test Method Development

Product and process understanding leads to identifying quality attributes that require analytical measurement for control, as described in a quality target product profile (QTPP). Measurement needs can be captured in an Analytical Target Profile (ATP) [6], which forms the basis for developing the analytical procedure. The ATP includes information on the intended purpose; product attributes to be measured, and relevant performance characteristics, as well as performance criteria. It also drives the choice of analytical technology. While multiple analytical techniques may meet the performance requirements, consideration of the operating environment should also be included in the selection. Once a technology is chosen, the ATP is a foundation for developing the appropriate analytical procedure attributes and acceptance criteria for analytical procedure validation. The Analytical Target Profile (ATP) helps in selecting the technology, designing and developing the procedure, and monitoring and improving the analytical procedure's performance over time. The ATP is kept updated throughout the procedure's lifecycle and can also serve as a foundation for managing its lifecycle (Fig. 2.8), ensuring that it remains suitable for its intended use.

Conducting experiments to investigate the range of parameters provide additional information about the analytical procedure's performance. The respective analytical

Hardness Sorted Parameter Estimates

Term	Estimate	Std Error	t Ratio	t Ratio	Prob>|t|
Impeller/Chopper Speed	-0.4	0.424853	-0.94		0.5192
Spray Rate	-0.2	0.424853	-0.47		0.7199
Kneading Time	-0.15	0.424853	-0.35		0.7839

Thickness Sorted Parameter Estimates

Term	Estimate	Std Error	t Ratio	t Ratio	Prob>|t|
Spray Rate	0.000425	0.000548	0.78		0.5800
Kneading Time	- 0.000325	0.000548	-0.59		0.6591
Impeller/Chopper Speed	0.000275	0.000548	0.50		0.7038

Friability Sorted Parameter Estimates

Term	Estimate	Std Error	t Ratio	t Ratio	Prob>|t|
Impeller/Chopper Speed	0.175	0.078262	2.24		0.2677
Spray Rate	0.075	0.078262	0.96		0.5135
Kneading Time	0.075	0.078262	0.96		0.5135

Assay Sorted Parameter Estimates

Term	Estimate	Std Error	t Ratio	t Ratio	Prob>|t|
Spray Rate	0.0375	0.124102	0.30		0.8132
Kneading Time	0.0375	0.124102	0.30		0.8132
Impeller/Chopper Speed	0.0125	0.124102	0.10		0.9361

Dissolution Study – Q Time Point Sorted Parameter Estimates

Term	Estimate	Std Error	t Ratio	t Ratio	Prob>|t|
Spray Rate	2	0.447214	4.47		0.1400
Impeller/Chopper Speed	1	0.447214	2.24		0.2677
Kneading Time	0	0.447214	0.00		1.0000

Fig. 2.7 Example summary of DoE: influence of CPP on CQA of tablets

procedure attributes and criteria are derived from the ATP. Examining a single parameter establish proven acceptable ranges (PAR) for the procedure. The approach involves studying the interactions and ranges of relevant parameters in multivariate experiments (DoE). Risk assessment and prior knowledge are used to identify the

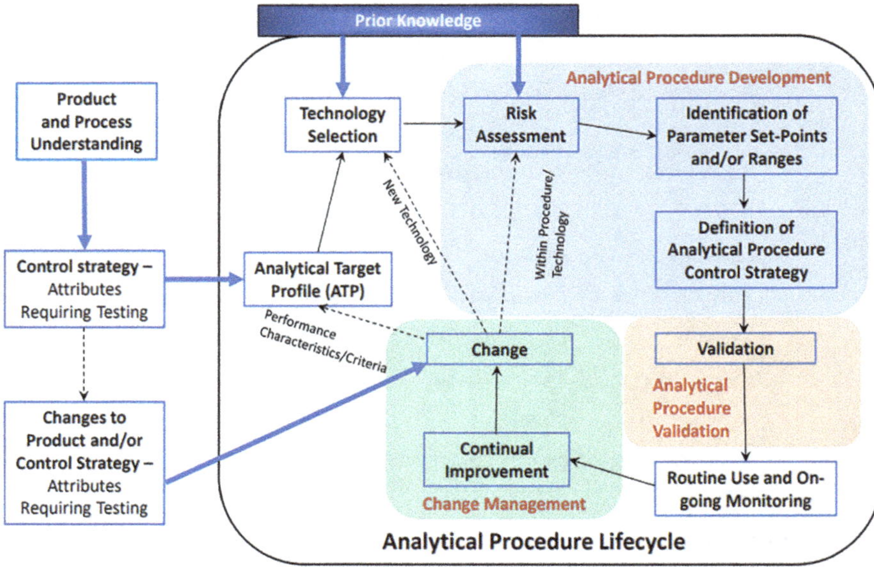

Fig. 2.8 Analytical procedure lifecycle. (Source: ICH Q14 [7])

parameters, attributes, and appropriate ranges to investigate. Categorical variables, such as different instruments, are considered in the experimental design. The results of these studies provide insight into the relationships between the analytical procedure's variables and its responses. Based on these results, fixed set-points are established for some parameters, PARs for others, and others may be included in a method operating design range (MODR). A MODR consists of combined ranges for two or more variables within which the analytical procedure is fit for its intended use. The analytical procedure needs to be validated for performance characteristics. An analytical procedure validation strategy, as part of the analytical procedure validation protocol, defines the necessary extent of validation.

Risk Assessment

A risk assessment (Fig. 2.9) is a systematic process for organizing and evaluating information to support a risk decision made within a quality risk management system. The process includes identifying hazards and analyzing and evaluating the risks associated with exposure to those hazards. In the pharmaceutical industry, a technical risk assessment includes risks associated with various aspects of the drug substance or product, including but not limited to early development, synthetic route development, drug substance process design, drug product formulation development, drug product process design, and process scale-up. The US FDA encourages using modern pharmaceutical development concepts and quality risk management at all stages of the manufacturing process lifecycle. Per the FDA

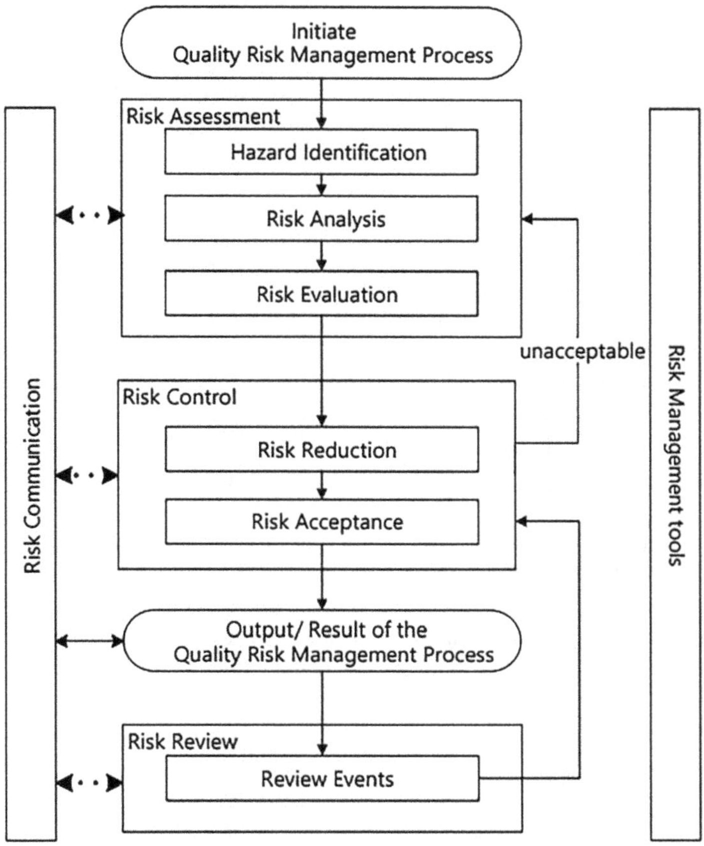

Fig. 2.9 QRM process. (Source: ICH Q9 (R1))

guidance on process validation, the lifecycle concept includes product and process development, qualification of the commercial manufacturing process, and maintenance of the process in a state of control during routine commercial production.

Establishing Control Strategy

Existing process understanding and prior knowledge of the scale-up process assist in defining scale-up plans, along with appropriate controls, to ensure consistent process performance and product quality. The controls include:

- Control of starting material attributes, in terms of critical quality attributes.
- Controls unit operations (critical process parameters).
- Testing of critical quality attributes of in-process material.
- Specifications on critical quality attributes of the finished product.
- Other controls identified.

Conclusion

In modern pharmaceutical formulation and process development, quality by design (QbD) principles are widely used to ensure that the final product consistently meets predefined quality attributes and specifications. The QbD approach involves understanding the relationship between critical product and process parameters and controlling those parameters to produce a high-quality product consistently. One key aspect of formulation and process development is the use of advanced analytical techniques to understand and control product quality. These techniques include spectroscopy, chromatography, microscopy, and thermal analysis, among others. Additionally, manufacturing processes often use advanced technologies such as continuous manufacturing to improve efficiency and reduce costs. Another important aspect of formulation and process development is using quality risk management (QRM) principles. QRM is a systematic process for identifying, evaluating, and controlling risks throughout the product development process, from research and development to manufacturing and distribution. Computer-aided technologies such as computational fluid dynamics (CFD) [8] and process simulation are increasingly being used to optimize the design of pharmaceutical manufacturing processes. It is important to understand that the knowledge acquired from early formulation and process development through scale-up will define what can be considered critical to control and an integral part of quality risk management (QRM). Overall, pharmaceutical formulation and process development is a complex process that requires a deep understanding of the science behind the product and the manufacturing process, as well as the use of advanced analytical techniques, advanced manufacturing technologies, and use of QbD and QRM principles.

References

1. Davidovich M, Mueller R, Raghavan K, Ranadive S, Vitez I, Sarsfield B (2001) The role of powder X-ray diffraction as a powerful tool in the characterization of various hydrates of drug substance. Am Pharm Rev 4:53–60
2. Byrn SR, Zografi G, X (Sean) Chen (2017) Differential scanning calorimetry and thermogravimetric analysis. In: Solid state properties of pharmaceutical materials. Wiley
3. FDA (2023) Inactive ingredients in approved drug products. US FDA
4. Qiu Y, Qiu Y, Chen Y, Zhang G, Chen Y, Zhang G, Liu L, Porter W (2008) Developing solid Oral dosage forms: pharmaceutical theory and practice. Academic
5. Politis SN, Colombo P, Colombo G, Rekkas DM (2017) Design of experiments (DoE) in pharmaceutical development. Drug Dev Ind Pharm 43(6):889–901
6. Barnett KL, McGregor PL, Martin GP, LeBlond DJ, Jane Weitzel ML, Ermer J, Walfish S, Nethercote P, Gratzl GS, Kovacs E (2016) Analytical target profile: structure and application throughout the analytical lifecycle. USP Stimuli
7. ICH (2022) ICH Q14 analytical procedure development. In: International council for harmonisation of technical requirements for pharmaceuticals for human use. ICH
8. Pordal HS, Matice CJ, Fry TJ (2002) The role of computational fluid dynamics in the pharmaceutical industry. Pharm Technol 26:72–77

Chapter 3
Quality Risk Management (QRM)

Abstract Risk assessment is a critical process that involves identifying, analyzing, and evaluating potential hazards and risks associated with the drug substance or product throughout the product development and manufacturing process. Quality by design (QbD) principles and quality risk management (QRM) are encouraged by the US FDA for consistent adherence to predefined quality attributes and specifications. QRM should be applied during various stages of product development to identify and mitigate potential risks, using technical risk assessment methodology to facilitate science-based decision-making in a systematic, objective, and efficient manner. Semi-quantitative risk assessment (SQRA) can help speed decision-making in fast-paced industrial environments by prioritizing projected product risks that may impact end users. In pharmaceutical manufacturing, it is often challenging to identify and define failures, making product and process risk more relevant and practical to assess. The use of software with an audit trail can efficiently update a risk assessment platform for a specific quality attribute, reducing the need for additional time for rework or repeat risk assessments for complex multiunit operations. Modern quality systems and risk assessment tools that utilize product and process knowledge can simplify regulatory approval requirements for post-approval changes to facilities, equipment, and processes.

Keywords Risk Assessment · Quality by Design (QbD) · Quality Risk Management (QRM) · Semi-Quantitative Risk Assessment (SQRA) · Pharmaceutical Manufacturing

Introduction

Risk assessment is a crucial step in ensuring the quality and safety of pharmaceutical products. It is a systematic process for identifying, analyzing, and evaluating potential hazards and risks associated with the drug substance or product. This includes evaluating risks related to various stages of product development, such as synthetic route development, drug substance process design, drug product formulation development, and scaling up the process. The US FDA encourages using quality by design (QbD) principles and quality risk management (QRM) throughout the

A. B. Pazhayattil et al., *Technology Transfer*, AAPS Introductions in the Pharmaceutical Sciences 10, https://doi.org/10.1007/978-3-031-32192-4_3

product development and manufacturing process. This approach ensures that the final product consistently meets predefined quality attributes and specifications and that risks are identified and managed throughout the product development process.

Quality risk management (QRM) is a vital step in ensuring the safety and effectiveness of pharmaceutical products. It is an integral part of the lifecycle, and it should be applied during various stages, including process design (Stage 1), process qualification (Stage 2), and continued process verification (Stage 3), to identify and mitigate potential risks. The purpose of technical risk assessment methodology is to facilitate science-based decision-making in a systematic, objective, and efficient manner, as recommended in the ICH Q9 Quality Risk Management guideline. QRM aims to improve the overall quality and safety of pharmaceutical products by identifying and managing risks effectively throughout the product development and manufacturing process.

Traditional QRM Tools

Quality risk management, as outlined in the ICH Q9 guideline, aims to protect patient safety and ensure that the level of risk management effort is commensurate with the level of risk. To achieve this, appropriate risk assessment tools must be chosen and employed during various stages of the product development lifecycle. While simple techniques such as flowcharts, fishbone diagrams, or check sheets can be used, more advanced tools such as Fault Tree Analysis (FTA), FMEA/FMECA, PHA, HACCP, and HAZOP [1] are also commonly employed in the risk assessment process. These tools may be more effective at specific stages of development. For example, a PHA tool may be best used when minimal data is available at early formulation development stages. It is important to ensure that the chosen tool is appropriate for the unique nature of the pharmaceutical process development requirements and product quality attributes. Risk assessment tools commonly use binary or ordinal scales such as FMEA Risk Priority Number (RPN) to measure the level of risk.

Most critical quality attributes (CQAs), such as assay and dissolution, are presented in numerical form. It is crucial for both patient safety and the efficiency of business operations to monitor and understand the trends in these CQAs. Therefore, it is beneficial to have a quantitative measure of the product's risk profile concerning CQAs. Hazard analysis and critical control points (HACCP) is a risk management approach that addresses physical, chemical, and biological hazards associated with the design, development, production, and use of the product. It is important to have risk assessment tools (Table 3.1) that are directly linked to patient safety when used appropriately.

Table 3.1 Risk assessments steps and proposed tools

Category	Steps requiring risk gauging	Departments/collaborators	Examples of recommended QRM tools[a]
1	Route selection	Research and Development (R&D)	Scientific rationale
	Impurity, stress studies	R&D/Analytical	Scientific rationale, Checklist
	Initial Quality by Design (QbD) process development	R&D	Fault Tree Analysis (FTA), Failure Mode and Effects (and Criticality) Analysis (FMEA/ FMECA)
	Occupational health toxicity/potency categorization	Toxicology/R&D, Technology Transfer	Preliminary Hazard Analysis (PHA), Hazards Analysis and Critical Control Points (HACCP)
	Registered Starting Materials (RSM) and key material supplier qualification	Quality/R&D	Ishikawa diagram, FMECA, Risk Ranking and Filtering
2	Post-QbD process development	R&D	Failure Mode and Effects (and Criticality) Analysis (FMEA/ FMECA), Semi-Quantitative Risk Assessment (SQRA)
	Pre-Analytical Quality by Design (AQbD) method development	Analytical	Fault Tree Analysis (FTA), Failure Mode and Effects (and Criticality) Analysis (FMEA/ FMECA)
	Pre scale up	R&D/Technology Transfer	Failure Mode and Effects (and Criticality) Analysis (FMEA/ FMECA), Semi-Quantitative Risk Assessment (SQRA)
	Facility/Equipment	Engineering/Technology Transfer	Failure Mode and Effects (and Criticality) Analysis (FMEA/ FMECA)
	Method transfer	Analytical/Quality	Failure Mode and Effects (and Criticality) Analysis (FMEA/ FMECA)
	Pre-clinical	Clinical/R&D	Checklist, Fault Tree Analysis (FTA), Hazards and Operability Studies (HAZOP)
	Stability study	Analytical/R&D	Fault Tree Analysis (FTA)
3	Drug Master File (DMF) submission	Regulatory Affairs (RA)/ Quality	Checklist
	Post scale up/pre-Process Performance Qualification (PPQ)	Technology Transfer/ Operations, Quality	Semi-Quantitative Risk Assessment (SQRA)
	Pre-Approval Inspection (PAI)	Quality/Technology Transfer, R&D	Checklist
4	Determination of Stage 2 PPQ batches	Technology Transfer/ Quality	Statistical tools to estimate the sample size and number of required PPQ batches

(continued)

Table 3.1 (continued)

Category	Steps requiring risk gauging	Departments/collaborators	Examples of recommended QRM tools[a]
	Post PPQ	Technology Transfer/ Quality	Semi-Quantitative Risk Assessment (SQRA)
	Determination of Stage 3a batches	Technology Transfer/ Quality	Statistical tools to estimate the sample size and number of required Stage 3a batches
5	Post Stage 3a	Technology Transfer/ Quality	Semi-Quantitative Risk Assessment (SQRA)
	Ongoing annual product review/Continued Process Verification (CPV)	Quality	Failure Mode and Effects (and Criticality) Analysis (FMEA/ FMECA)
	Investigations	Quality	Ishikawa diagram
	Change management	Quality	Ishikawa diagram

[a]The column provides a list of potential application tools only; other tools may be used at these stages based on user preference and applicability

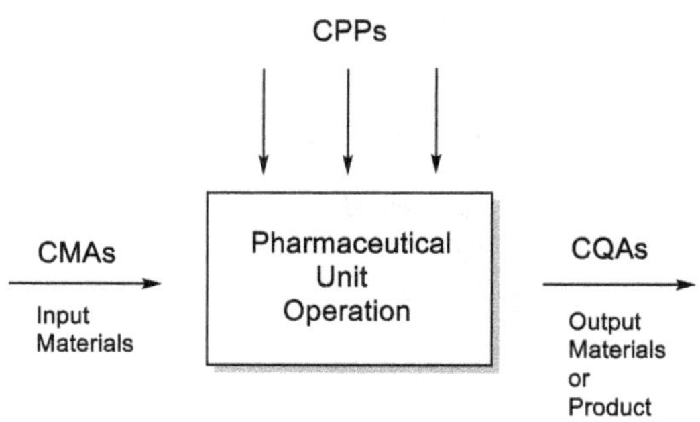

Fig. 3.1 Relationship between CMAs, CPPs, and CQAs

Semi-Quantitative Risk Assessment (SQRA) Methodology

The semi-quantitative risk assessment (SQRA) methodology [2] is designed to address these two important considerations. Quality risk management is a systematic process for assessing, controlling, communicating, and reviewing risks to the quality of the drug product throughout its lifecycle. SQRA is a systematic and data-driven risk assessment tool that can be utilized before and after sufficient product/process design of experiments (DoE) data has been generated. It provides a quantitative measure of the risk profile of the product and links it directly to patient safety.

The performance of a specific critical quality attribute (CQA) is influenced by a range of critical material attributes (CMAs) and critical process parameters (CPPs) that have the potential to impact the CQA results significantly. The relationship between CMAs, CPPs, and CQAs is depicted in Fig. 3.1 and is the fundamental principle behind the design of the semi-quantitative risk assessment (SQRA) methodology.

Risk assessment is a process that includes identifying, analyzing, and evaluating risks. The semi-quantitative risk assessment (SQRA) methodology incorporates distinct steps tailored to prioritize patient safety as the goal. It employs an innovative, reliable, robust, objective, and data-driven approach to risk estimation. The semi-quantitative risk assessment (SQRA) methodology includes the following steps:

1. Define the scope of the risk assessment and review relevant product/process information.
2. Identify a critical quality attribute (CQA) of the product manufactured by the process.
3. Identify a plurality of material attributes (MAs) (e.g., API particle size, bulk density) and/or a plurality of process parameters (PPs) (e.g., compression force, press speed) that are considered to have a potential impact on the CQA.
4. Classify each of the MAs and/or PPs into a risk category using a decision tree for risk category determination.
5. Determine the overall risk score and ratio for the CQA based on risk factors designated for each of the risk categories and count the MAs and/or PPs in each of the risk categories.
6. Visualize the outcome of the SQRA using various types of charts.

This methodology is designed to provide a quantitative measure of the risk profile of the product and link it directly to patient safety. It is a robust, reliable, objective, and data-driven approach to risk estimation, which is customized to prioritize patient safety (Fig. 3.2). SQRA methodology can be applied to any type of production, including batch production, continuous process, or a combination of the two, at any stage of the product lifecycle. This includes early development, pilot formulation development, or commercial manufacturing process, to evaluate risks associated with the product, process, scale-up, and changes in the product lifecycle.

The SQRA can be conducted at any stage of the product lifecycle, such as early development, formulation development, or commercial manufacturing process. The stage of the manufacturing process lifecycle that the current SQRA is relevant to must be clearly defined to meet the corresponding regulatory framework and expectations. For example, FDA/TPD/EMA requirements for various drug applications and subsequent submission activities (such as scale-up and post-approval changes) or any international guidance (ICH, WHO) on pharmaceutical manufacturing/quality system practice.

Define Scope and Review Relevant Product/Process Information

The first step in the semi-quantitative risk assessment (SQRA) methodology is to define the scope of the risk assessment and review relevant product/process information. During the Scope determination and Product/Process review stage, the objective and scope of the current assessment are identified and confirmed with

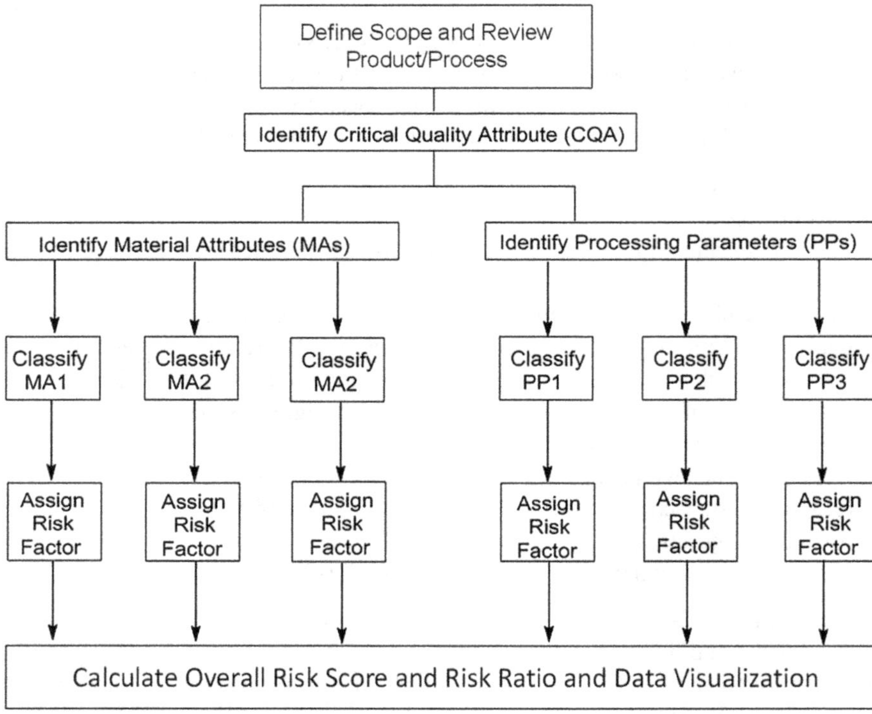

Fig. 3.2 Semi-quantitative risk assessment (SQRA) process steps

project owners and stakeholders. A process flow map or fishbone diagram can be constructed and modified as needed to cover all operations and controls in the process under evaluation. For products developed without a quality by design (QbD) approach, a gap assessment can be used to determine if additional commercial data collection is needed before the assessment is carried out. The data can then be used to determine the CMAs and CPPs that have a direct impact on non-label and label-claimed CQAs.

CQA Identification

Attributes refer to the physical, chemical, biological, or microbiological properties or characteristics of an in-process or finished product. Quality attributes can be divided into two categories: those that directly impact the end user, known as label claim quality attributes, and those that do not have a direct impact on the end user, known as non-label claim quality attributes. Non-label claim quality attributes may still have an indirect effect on label claim quality attributes. The SQRA process applies distinct decision-making criteria for each category. Examples of label claim and non-label claim quality attributes are provided in Table 3.2.

Table 3.2 Quality attribute examples

Category of quality attributes	Examples (may vary from product to product)
Label claim quality attributes	Assay, Concentration, Content Uniformity, Dissolution, Drug Release, Fill Weight, , Sterility Assurance Level, Potency
Non-label claim quality attributes	Adhesion, Assay Per Spray, Blend Uniformity, Bulk Density, Cold Flow, Disintegration, Droplet Size, Extractable Volume, Friability, Hardness, Integrity, Osmolality, Particle Count, Particle Size Distribution, Particulate Matter, Peel, pH, Sheer, Specific Gravity, Spray Pattern, Tack, , Torque, Unit Weight, Viscosity, or Weight Variation

Table 3.3 Material attribute, process parameter examples

Category	Examples
MA	API – Particle size, stability, solubility, BCS class Excipient – quantity, grade, viscosity, particle size, surface morphology/surface area, density, impurities/degradation products
PP	Solid Dosage form – Compaction force, screen size, roller speed, roller type, roller gap, compression speed/force, coating gun spray rate, coating pan speed, coating pan temperature Liquid Dosage form – mixing speed/time, temperature, compounding time, hold time, tubing type, filer type, purge volume/time, fill volume.

Material Attribute (MA)/Process Parameter (PP) Identification

To identify potential material attributes (MAs) such as drug substance and excipients and process parameters (PPs) that may impact critical quality attributes (CQAs) (Table 3.3) during the risk assessment process, a process flow map or fishbone diagram from the formulation process and/or optimization process can be used as a guide. Identifying the appropriate MAs and PPs for risk assessment is a collaborative effort involving a team of subject matter experts. Below is a brief list of examples of MAs and PPs for both solid and liquid dosage forms.

MAs/PPs Classification and Risk Factor Assignment

As per ICH Q9 guidance, risk analysis is the process of linking the probability of occurrence, the severity of harm, and the detectability of the hazard (depending on the tool used) associated with the product/process. Determining these terms often requires extensive information collection and can be very subjective, based on various perceptions of different functional groups. Numerical values or surrogate descriptors for probability, severity, and detectability are usually assigned as discrete ordinal numbers or qualitative descriptors (low, high). It is important to note that the magnitude of these numbers does not necessarily mean it to be on an exact numerically proportional scale. An event with a probability number of 2 does not necessarily translate into a probability that is twice as likely to occur as an event with a probability number of 1. Additionally, it is mathematically flawed to multiply ordinal scale numbers [3]. The validity of such multiplication (e.g., the risk number = 2 × 3 × 4) has been debated for some time. To overcome this issue, in the SQRA methodology, a hybrid numbering system of ordinal risk categories that represents the combination of probability and severity is created. It can be presented in the form of a decision tree. All MAs and PPs subject to the risk assessment process will be examined against the decision tree (Figs. 3.3 and 3.4) and assigned an ordinal

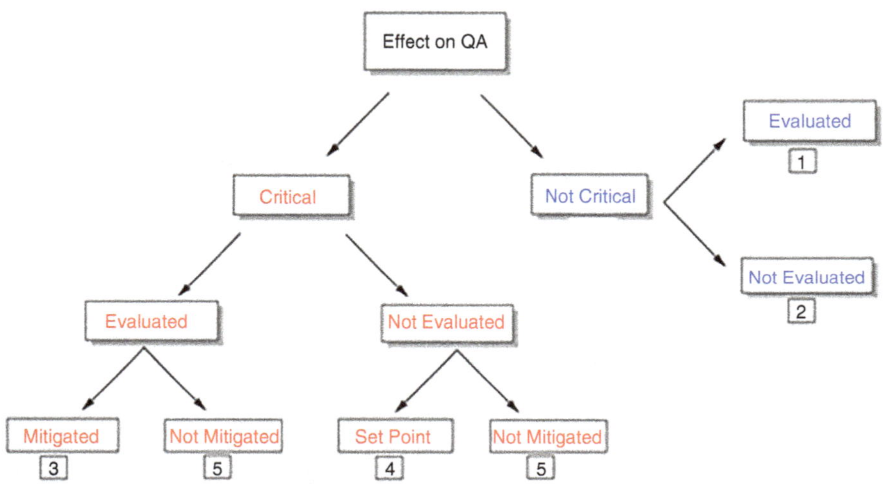

Fig. 3.3 Quality attributes with no direct end-user impact

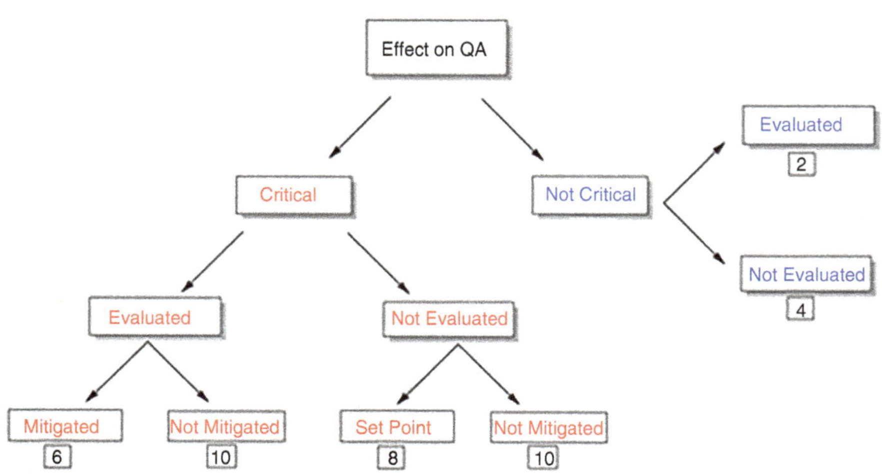

Fig. 3.4 Quality attributes with direct end-user impact

Table 3.4 Definition of each risk category

Risk category	Risk factor-for MA/PP relating to non-label claim quality attributes	Risk factor-for MA/PP relating to label claim quality attributes
Not critical evaluated	1	2
Not critical-not evaluated	2	4
Critical-evaluated and mitigated	3	6
Critical-not evaluated with set point	4	8
Critical-not mitigated (either evaluated or not evaluated)	5	10

risk category accordingly. This allows for a more accurate and meaningful assessment of the risk associated with the product/process and supports risk-based decision-making. The initial decision tree is utilized for MAs and PPs associated with non-label Claim quality attributes, which do not have a direct impact on patients. The second decision tree is applied to those related to label claim quality attributes that have a direct impact on patients, such as assay, content uniformity, and dissolution. Each category is assigned a risk factor that is double that of the corresponding categories from the non-label claim quality attributes, with values of 2, 4, 6, 8, or 10 (as per Table 3.4).

To effectively classify each identified MA, PP, and their interactions into the appropriate risk category, it is crucial for the project owner and SMEs to engage in collaborative, data-driven decision-making. Utilizing product history, such as results from early development trials, can provide valuable insights. Additionally, informed opinions based on literature review and previous experience with similar products/processes can be beneficial. A key feature of these risk factors is that they are established primarily based on data, as opposed to conventional probability and severity numbers. The first step in this process is to use science-based decision-making to categorize the MA/PP of interest as critical or not critical. Then, the MA/PP of interest can be further divided into subgroups within these two main groups based on whether or not it has been evaluated and mitigated.

An important aspect of the alternate risk assessment is that it views the probability of occurrence and the severity of the consequence as a dynamic combination, rather than two separate, static factors. For example, varying a critical PP-like compression force without proper control can significantly impact tablet hardness, resulting in a risk factor of 5 if no supporting data is available from the process. However, by implementing a mitigation plan, such as studying the effect of compression force on tablet hardness and proposing a proper operating range, the risk factor can be adjusted to 3. The SQRA risk factor provides an objective measure of the overall risk profile. Additionally, the conventional concept of detectability is also reflected in the alternate SQRA risk factor system. Specifically, MAs that are evaluated with preexisting specifications and testing requirements should be easily detected and quantified in a validated measurement system. However, for MAs and PPs that have not been evaluated for the specific product/process, or for which similar product/process data is used, there may be a question of detectability, which is reflected in the relatively higher risk factors associated with them.

Calculating Risk Ratio

Once all the CMAs and CPPs have been assigned risk factors, the overall risk score and risk ratio of the process/product can be determined using validated data compilation and software systems. The risk score for each category is calculated by multiplying the total number of MA/PP factors per category with the assigned risk factor. A higher risk score indicates a greater level of risk for that CQA within the process under review. The overall risk score is the sum of the risk scores for all categories. The value of the risk score is directly linked to the number of MAs/PPs included in the risk assessment. Therefore, to make meaningful comparisons, the risk scores should be calculated using the same number of total MAs/PPs. For example, risk scores for the same CQA of the same product from different product lifecycle stages (formulation development vs. process validation), or for the same CQA of two similar products at a specific product lifecycle stage (higher strength vs. lower strength).

Unlike the risk score, which cannot be interpreted in isolation/out of context, the risk ratio is a more independent indicator that can be compared among different situations. The overall risk ratio is computed as the sum of the count of CMAs and/or CPPs under categories "critical not evaluated with set point" and "critical not mitigated" divided by the sum of the count of CMAs and/or CPPs under categories "critical mitigated," "critical not evaluated with set point," and "critical not mitigated." In another word, the risk ratio represents the portion of the CMAs/CPPs that are not evaluated and/or not mitigated out of the total CMAs/CPPs that are identified. The risk ratio is calculated from the following equation.

$$\text{Risk Ratio} = \frac{\begin{array}{c}\text{Sum of Count of CMA/CPP Factors (Critical/Not Evaluated}\\ \text{with Set Point} + \text{Critical/Not Mitigated)}\end{array}}{\begin{array}{c}\text{Sum of Count of CMA/CPP Factors (Critical/Mitigated} + \text{Critical/}\\ \text{Not Evaluated with Set Point} + \text{Critical/Not Mitigated)}\end{array}}$$

The risk ratio is a fraction between 0 and 1 that expresses the relative risk of an event occurring. It can be used to establish critical limits based on a company's risk tolerance, which considers both strategic considerations and regulatory requirements. An example of this can be seen in Table 3.5, which shows how risk ratios can be classified into different levels. Organizations can set their own threshold values as needed.

Table 3.5 Risk threshold limit example

Risk ratio	Risk
Less than 0.25	Low risk
0.25–0.75	Medium risk
Higher than 0.75	High risk

Category	Count of CMA/CPP Factors in Each Category	Risk Factor	Risk Score for Each Category
Not Critical	0	0	0×0 = 0
Not Critical/Evaluated	31	1	31×1 = 31
Not Critical/Not Evaluated	26	2	26×2 = 52
Critical/Mitigated	41	3	41×3 = 123
Critical/Not Evaluated with Set Point	16	4	16×4 = 64
Critical/Not Mitigated	2	5	2×5 = 10
Total Risk Score	31+52+123+64+10 = 280		
Risk Ratio	(16+2)/(41+16+2) = 0.31		

Fig. 3.5 Example SQRA technical risk assessment during stage 1 (Process design)

Risk Assessment Chart

Using the risk ratio calculated in Fig. 3.5, risk assessment charts can be created for critical quality attributes (CQAs) using monitoring and measuring activities (MAs) and/or process parameters (PPs). These charts should indicate the risk category and level of significance for each MA and/or PP. A common way to display this information is through a heat map (Fig. 3.6) or a Pareto chart. Other charts that can be used include bar charts, bubble charts, cone charts, pie charts, radar charts, scatter charts, surface charts, and prioritization matrices.

Based on the heat map and risk ratio for each stage, the most critical monitoring and measuring activities (CMAs) and critical process parameters (CPPs) can be identified and optimized to minimize or eliminate the identified risks. After optimization, the technical risk assessment (SQRA) should be repeated until the risk ratio is reduced to medium or low risk. Since product development is an iterative process that involves multiple risk assessments, it is important to have adequate access and traceability of SQRA reports for effective product knowledge management and the application of a lifecycle approach. Any recommendations for mitigation actions and post-SQRA risk assessments should be documented within the firm's quality management system (QMS).

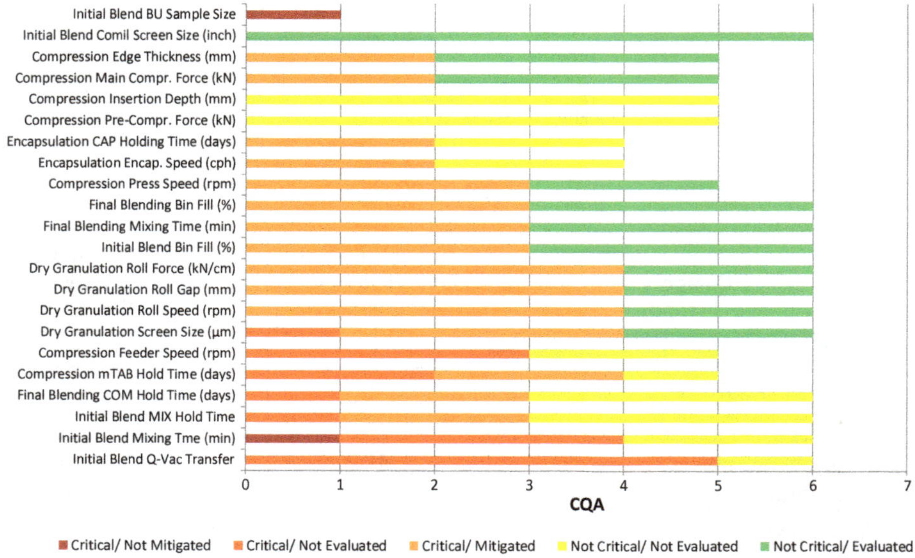

Fig. 3.6 Example scale-up SQRA heatmap: CPP vs CQA

Comparison

Below is a comparison of the SQRA approach to conventional risk assessment tools (Table 3.6).

Relevance of Data-Driven Tools

The revised guideline (ICH Q9 (R1)) acknowledges that some level of subjectivity in quality risk management (QRM) [4] is inevitable and recommends reducing or controlling it by using QRM tools properly and utilizing relevant data and knowledge sources. All personnel involved in QRM are expected to be aware of, anticipate, and address the potential for subjectivity, according to the revised guidance. Additionally, decision-makers should ensure that subjectivity is controlled and minimized to support more robust risk-based decision-making.

The later Q9 also includes a new section on the concept of formality in quality risk management (QRM). The ICH states that there is a lack of understanding of what constitutes formality in QRM and how it can be further developed to improve the application of QRM principles. The guideline suggests that an understanding of formality in QRM can lead to more efficient use of resources. Lower risk issues can be handled with less formal means, allowing more resources to be dedicated to managing higher risk issues and more complex problems that may require more rigor and effort.

Table 3.6 A general comparison table

Facet	SQRA	FMEA/FMECA	FTA	PHA	HACCP
Risk focus	Deviation of CQAs that can link to end-user safety	Failure modes	Process faults/deviations	Harm (hazards, faults)	Hazards
Tool approach	Top-down approach; identify factors that impact the CQA and their extent of the impact	Bottom-up approach	Top-down approach; consider what causes a failure	Bottom-up approach	Bottom-up approach
Input sources	Historical data, SME technical input or product and process data	Historical data, SME technical input or product and process knowledge	Output from other hazard analysis; comprehensive process and cause-effect understanding	Highly dependent on SME input	Comprehensive understanding of processes and controls
Outcome indicator	Semi-Quantitative – Risk score and risk ratio	Semi-Quantitative – RPN	Qualitative – failure path and estimated probability	Qualitative risk rating	Qualitative
Risk ranking	Yes, by ordinal numbers	Yes, by ordinal numbers	Yes	Yes, semi-quantitative	By risk category (critical/noncritical)

Conclusion

Quantified risk profiles like those developed using SQRA can help to speed up decision-making in fast-paced industrial environments. Unlike failure/hazard analysis, SQRA prioritizes projected product risks that may impact end-users, rather than focusing solely on failures or hazards. In pharmaceutical manufacturing, it is often difficult to identify and define failures, failure modes, or hazards, making product and process risk more relevant and practical to define and assess. Once a risk assessment platform for a specific quality attribute is established, the risk assessment can be updated continuously and efficiently using software with an audit trail, reducing the need for additional time for rework or repeat risk assessments for complex multiunit operations. Regulators have determined that using modern quality systems, along with risk assessment tools that utilize product and process knowledge, can simplify regulatory approval requirements for post-approval changes to facilities, equipment, and processes.

References

1. PDA (2012) TR 54 implementation of quality risk management for pharmaceutical and biotechnology manufacturing operations. Parenteral Drug Association
2. Pazhayattil AB, Sayeed-Desta N, Chen S, Ingram M (2020) A Semiquantitative risk assessment methodology fit for biopharmaceutical life cycle stages. PDA J Pharm Sci Technol 74(4): 423–434
3. Strong JC, McDermott TS, Oliver Heinzerling H, Claycamp G (2017) The risk of trusting risk priority numbers. AAPS Newsl 20:14–18
4. Murray KS, Reich S (2011) Quality Risk Management (QRM) tool selection: getting to right first time. Pharmaceutical Engineering

Chapter 4
Technology Transfer Process

Abstract Technology transfer (TT) is a crucial step in the pharmaceutical industry to ensure that the documented knowledge, experience, and proven ability of a sending unit are effectively transferred to a receiving unit. TT can include the transfer of manufacturing, packaging processes, and analytical testing from one site to another. Analytical transfer is the process of transferring analytical testing from one site to another. The successful transfer of analytical methods is critical to ensure the consistency, quality, and safety of pharmaceutical products. TT can be challenging due to the complexity of the processes, equipment, and the criticality of quality attributes. However, effective TT can result in cost savings, efficient manufacturing, and a more robust supply chain. Commercial product technology transfer is one type of TT that happens when a product is already in commercial production, and the development work, scale-up, process validation, and regulatory approvals have all been completed by the sending unit. The receiving unit is the production site that receives the product, process, and analytical methods. The successful completion of TT requires a robust, systematic approach with clearly defined roles and responsibilities, adequate training, effective communication, and collaboration between all parties involved. TT is subject to regulatory scrutiny and must meet regulatory requirements to ensure patient safety and compliance.

Keywords Technology Transfer · Analytical Transfer · Commercial Product Transfer · Manufacturing · Quality

Introduction

Technology transfer is the systematic procedure followed in passing the documented knowledge and experience gained during the development and/or commercialization to an appropriate, responsible, and authorized party. It embodies the transfer of knowledge, documentation, and the demonstrated ability of a receiving unit to effectively perform the critical elements of the transferred technology to the satisfaction of all parties and regulatory bodies. The analytical transfer is the process of transferring analytical testing from one site to another. Technology transfer

A. B. Pazhayattil et al., *Technology Transfer*, AAPS Introductions in the Pharmaceutical Sciences 10, https://doi.org/10.1007/978-3-031-32192-4_4

encompasses the process of transferring the manufacturing, packaging process, and analytical testing from one site to another.

$$\text{Tech transfer (TT)} = \text{Process transfer} + \text{Analytical transfer}$$

Commercial product technology transfer [1] happens when the product is already in commercial production. The development work, scale-up, process validation, and regulatory approvals have all been completed by the sending unit (SU). The receiving unit is defined as the production site that receives the product, process, and analytical methods. Sending unit is defined as the production/ development site with the approval to manufacture, package, and test the product.

Technology transfer can happen in the following different scenarios (Fig. 4.1).

Product transfers should be compliant with cGMP guidelines, streamlined, and on schedule. To achieve this, it should follow approved standard operating procedures, maintain a process-oriented and organized approach (Fig. 4.2), and have a flexible structure with dedicated teams.

Technology Transfer Phases

Per the WHO guidance [2], the technology transfer project plan is divided into four phases:

- Phase I: Project initiation
- Phase II: Project proposal (including forming a team, risk assessment, and creating a project plan)
- Phase III: Project transfer
- Phase IV: Project review

Phase I: Project Initiation

During the initiation phase of the project, a unit should identify the need for technology transfer, which may be due to lack of capacity, transfer from development to commercial site or transfer from one company to another, and establish initial discussion to identify if there is any interest for such a project, also considering the RU's ability to accommodate the intended activity, having the necessary technical expertise, technology, and capability, and a sufficient level and depth of detail to support the activity, and any further development and optimization at the RU.

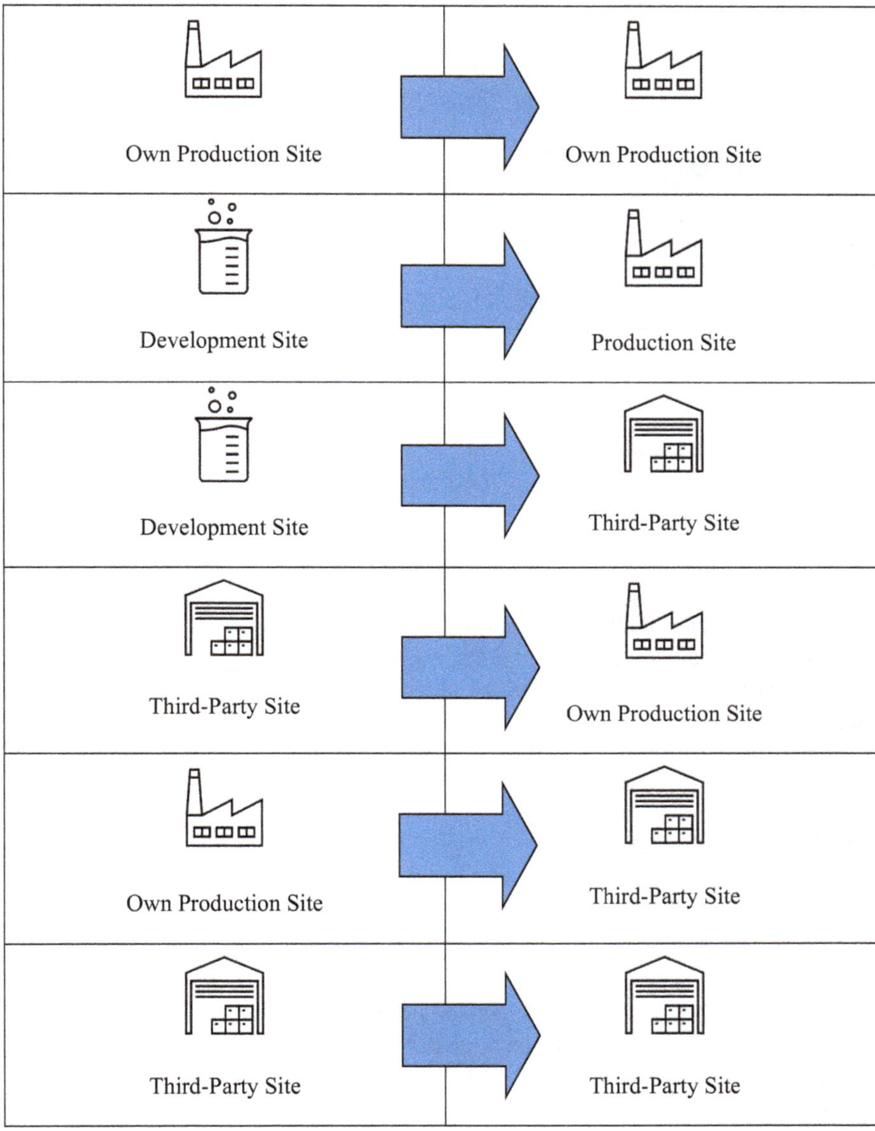

Fig. 4.1 TT scenarios

Phase II: Project Proposal

The SU and RU should establish a joint team to coordinate activities and execute the technology transfer exercise, perform a risk assessment based on the available data, information and knowledge of the premises, materials, products, procedures, and other related information, prepare the technology transfer document, develop a

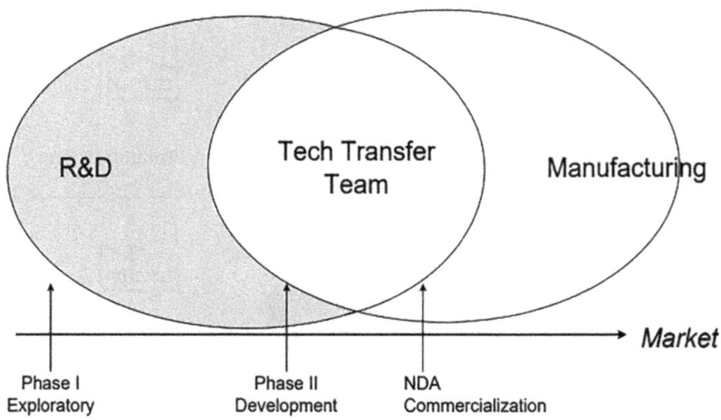

Fig. 4.2 Partnership philosophy

control strategy that includes risks, material attributes, processing steps and stages in production, testing steps in QC, equipment working principles and their impact on the process, critical quality attributes (CQAs), critical process parameters (CPPs), and in-process controls, QC instruments, acceptance criteria and limits, alarms and trends, personnel requirements such as qualification and training and qualification and validation.

Phase III: Project Transfer

The team should ensure that the starting materials (Active Pharmaceutical Ingredients (APIs) and excipients) used at the RU are consistent with those used at the SU regarding specifications and relevant functional characteristics. Any properties impacting the process or final product should be identified and characterized. The SU should provide the RU with the open part of the Drug Master File (API master file) or equivalent information, as well as any relevant additional information on the API of importance for the manufacture of the pharmaceutical product. This information should include, but is not limited to, the manufacturer and associated supply chain, the step of the API to be transferred, a flow chart of the synthesis pathway outlining the process, including entry points for raw materials, critical steps, process controls and intermediates, the definitive physical form of the API and any polymorphic and solvate forms, solubility profile, pH in a solution, partition coefficient, intrinsic dissolution rate, particle size and distribution, bulk physical properties, water content and determination of hygroscopicity, microbiological considerations, specifications and justification for release and end-of-life limits, stability study conclusions and recommendations, list of potential and observed synthetic impurities and degradants, potency factor, and any special considerations with implications for storage and/or handling, including safety and environmental factors and sensitivity to heat, light, or moisture. In summary, the SU should provide the RU with detailed information on the active pharmaceutical ingredients (APIs) and excipients to be used in the production of the pharmaceutical product. This information should include information on the manufacturers and supply chains, as well as specific data on the properties and characteristics of the materials, such as solubility, particle size, pH, and stability. Additionally, information on any potential impurities or degradants should be provided, along with any special considerations for storage and handling. Finally, the SU should provide regulatory documentation to support compliance with relevant certifications, such as TSE certification.

The transfer process from a SU to an RU should include the transfer of essential knowledge on the product, process, and procedures. This includes the quality target product profile, critical quality attributes, critical process parameters, material attributes, control strategy, and other elements that impact the product's quality. The SU should provide information on the product's history of development, scale-up activities, full-scale development, change history, and any health, safety, and environmental issues associated with the manufacturing process. The RU should identify any differences in facilities, systems, and capabilities and address them to assure equivalent product quality and develop relevant site operating procedures and documentation before starting routine production. The RU should also conduct process development tasks such as comparison and assessment of facility and equipment, determination of critical steps in manufacture, writing and approval of SOPs, evaluation of stability information, and compliance with regulatory requirements.

The transfer of packaging and analytical method operations should follow established procedural principles and include specifications for suitable containers and closure systems, as well as any relevant additional information on design, packing, processing, or labeling requirements and tamper-evident and anticounterfeiting measures. The RU should perform quality control testing of packaging components and analytical methods before testing samples for process validation studies. A protocol should be prepared for the transfer of analytical methods, including a description of the objective, scope, and responsibilities of the SU and RU, specifications of materials and methods, experimental design and acceptance criteria, documentation, the procedure for handling deviations, and details of reference samples. The SU should provide method-specific training, assist in the analysis of QC testing results, and provide validation reports and reference samples. The RU should ensure that necessary equipment and trained personnel are in place, execute the transfer protocol, perform validation, and generate and approve transfer reports. Training should be provided and documented, and reference should be made to compendial monographs. An experimental design should be prepared with acceptance criteria for the main analytical testing procedures. All relevant analytical procedure development and validation documentation should be made available by the SU to the RU.

Cleaning procedures and their validation should be site-specific. The SU should provide information on cleaning at the SU to minimize cross-contamination, including information on cleaning agents used, cleaning validation reports, and recovery studies to validate the sampling methodology. The SU should also provide information on limits for product residues and the rationale for limit selection. The RU should then design cleaning procedures considering relevant characteristics of the starting materials, manufacturing equipment design and configuration, cleaning agents, and product residue.

Phase IV: Project Review

Technology transfer should be monitored and reviewed throughout the project and evaluated upon completion to ensure that procedures and protocols are followed. Deviations or changes are properly documented and investigated. Statistical calculations and evaluations should be performed on the data and results, where possible, to identify trends, ensure compliance with control limits, and conduct capability studies. A comprehensive technology transfer report should be created based on the information and data obtained throughout the project and supported with relevant data that is easily accessible. The report should be authorized by the appropriate individuals and include a conclusion that summarizes the findings and outcome of the technology transfer.

For practical application, the fundamental components of technology transfer can be summarized into planning, transfer, and post-transfer.

Planning

In the planning, a transfer team is appointed, responsibilities are assigned, key information is exchanged between the transferor and recipient site, regulatory requirements are evaluated, and readiness assessments are conducted. Additionally, checklists for activities and deliverables are created, training activities are planned, and a Tech Transfer Plan is established, which includes a time frame and success evaluation criteria. The Technical Package (TP) is also assembled and approved during this phase. The key milestones include approval of the Technical Package and recipient site readiness for the transfer.

The transfer team should consist of the project-responsible personnel from both the sending unit (SU) and the receiving (launch) unit (RU), analytical experts from both sites, a project manager, a logistic manager from the RU, regulatory personnel, and coordinators. The team should clearly define responsibilities, including the one mainly responsible for the transfer, the one main contact person at each site, and a core team.

The SU provides the entire documentation to the RU, responsible for setting up local documents based on the Technical Package (TP) and Standard Operating Procedures (SOPs) in place. The transfer team should initiate elaborating a detailed transfer plan [3] (Fig. 4.3), including all activities and timelines (transfer checklist), as early as possible.

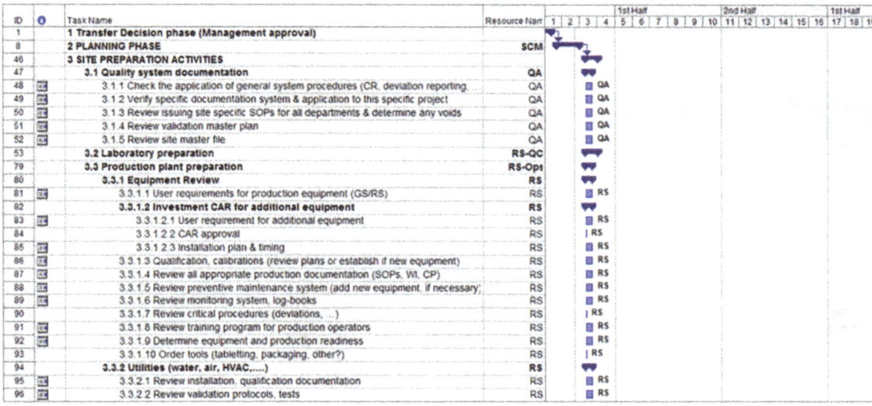

Fig. 4.3 Tech transfer plan

The team should also check for any obstacles for the intended purposes and meet to transfer difficult products or products with complexity. The RU is responsible for the logistics of all steps in the transfer process, including approvals, providing qualified and calibrated analytical equipment, reagents, and standards, and purchasing and releasing API and raw and packing materials. The RU is also responsible for checking whether cleaning for the new molecule/dosage form is already covered by cleaning validation performed at the production site and adapting as necessary. Risk assessment and documenting actions should also be performed.

Transfer

The transfer includes the actual transfer of technology and the implementation of the Tech Transfer Plan. The transfer phase of technology transfer includes the transfer of analytical methods and the process, a transfer assessment, the creation of a Transfer Report, and the identification of follow-up actions, with the key milestone being the approval of the Transfer Report.

Analytical Method Transfer has two transfer (Fig. 4.4) strategies: Handover Strategy, which is useful for transferring to external or geographically remote labs but requires detailed documentation and has a high potential for difficulties, and Partnership Strategy, which involves resolving difficulties jointly, conducting joint validation/precision/specificity studies and combined efforts which results in reduced issues.

Documents and protocols are essential for a successful technology transfer, and based on the Technical Package (TP), the receiving unit (RU) should set up a local set of documents, including a Technology Transfer Protocol/Report, control procedures for Active Pharmaceutical Ingredient (API), excipients, packaging

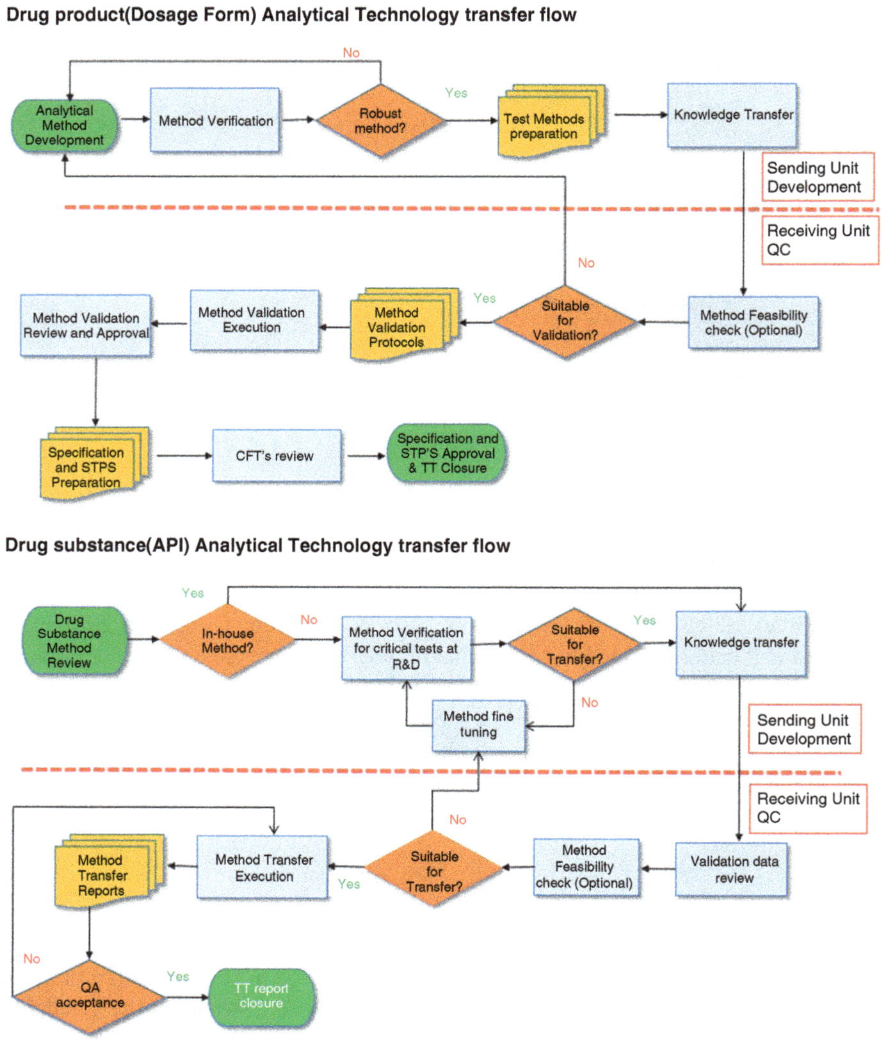

Fig. 4.4 Analytical transfer process

materials, finished dosage form (FDF), production instruction (BMR/BPR), process validation protocol, stability protocol. RU releases API, excipients, and packaging materials according to the approved control procedures and specifications. In case RU is not equipped to perform specific tests, it is responsible for outsourcing, following general Good Manufacturing Practices (GMP) rules [4] for the outsourcing of release tests. RU must ensure site readiness before starting manufacturing activities, including the availability of rooms and equipment (with qualification and calibration), environmental conditions and all raw materials, and

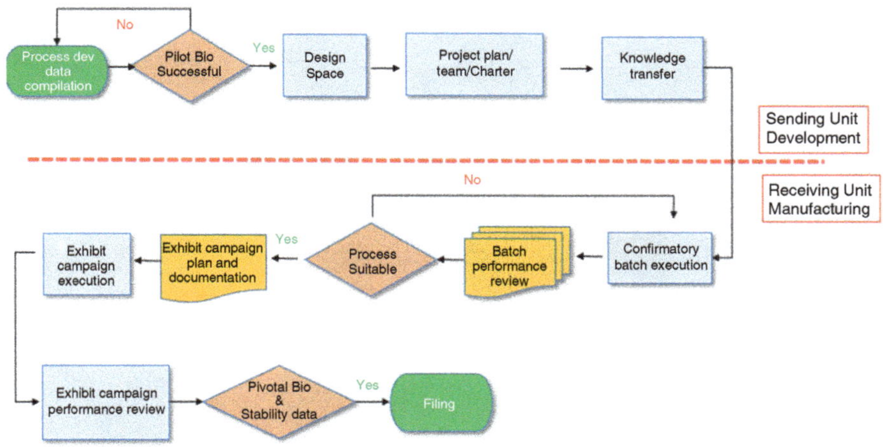

Fig. 4.5 Manufacturing process transfer flow

documents. The manufacturing process transfer (Fig. 4.5) should be done in cooperation with experts from sending unit (SU) and RU according to the master formula. Quality check is critical. Samples of FDF must be analyzed by QC of RU for batch release and by the analytical lab of SU for comparison, and only in case of matching results can the batch be released, and the transfer approved. Otherwise, investigations are necessary.

The stability testing of the transfer batches is the responsibility of the receiving unit (RU). It should be done according to the rules for stability testing of production launch batches and the stability protocol, with the response functions of the sending unit (SU) and RU reviewing the reports and results of all transfer activities and jointly approving a successful technical transfer, requiring a formal approval document signed by both sites and their respective quality assurance functions that includes a summary report.

Post-transfer

The post-transfer includes the evaluation of the success of the transfer and any necessary follow-up actions. The post-transfer phase of technology transfer involves monitoring the process during ongoing production, controlling and timely implementing follow-up activities such as stability evaluation, regulatory activities, customer activities, and termination of the transfer, with the key milestone being the completion of all transfer activities and transfer closure. The transfer can be considered successfully closed if the recipient site (RU) can routinely reproduce the transferred method, process, or product against a prescribed set of specifications as

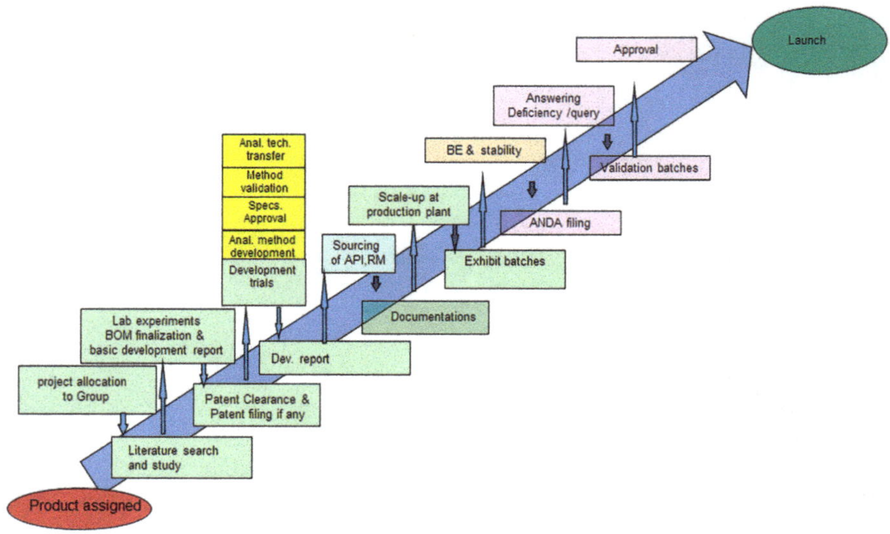

Fig. 4.6 R&D to site transfer

agreed by the transfer team and if the RU can demonstrate through clear documentation that the regulatory requirements and the necessary business needs have been satisfied.

R&D to Site Transfer

The transition from R&D to manufacturing (Fig. 4.6) involves the transfer of formula and process rationale, the evolution, and scaling up of analytical and manufacturing technologies, data, and experience from pivotal lots, the history of clinical batches, documentation capturing knowledge, the rationale for proposed specifications, and stability data.

During the assessment and evaluation, the SU's R&D group leader, scientist, or CRO scientist will be responsible for gathering and providing information on the manufacturing process to the receiving unit. A general process flow chart for the technology transfer can be found:

Product Development at R&D/Contract Research Organization

⇓

Sharing of TT Documents by R&D/CRO (SU) to RU

⇓

Planning of Trial / Scale Up / Exhibit Batches at RU
(Indenting for RM / PM / Tooling / Change Parts, License Applications to FDA etc.)

⇓

Initiation of scale-up review checklist by RU
Preparation and Issuance of documents for trial / scale-up batches (Sampling protocol, Specifications, STP, MFC, BMR, BPR, Cleaning method etc.)
Analytical methods transferred by RU
Analytical method transfers by R&D/CRO

⇓

Execution of Trial / Scale up batch(es)

⇓

Evaluation of Trial / Scale-up batch(es), scale-up report (as applicable) by RU and Sharing of TT documents by R&D/CRO to RU, as applicable.

⇓

Initiation of Checklist for exhibit batch start-up (Annexure-3) by RU MS&T
Preparation and issuance of documents for exhibit batch(s) (BMR, Process Qualification / Validation Protocol

⇓

Execution of exhibit batch(es)

⇓

Stability study of exhibit batch(es)

⇓

Evaluation of exhibit batch(es) (Acceptance of Product Process technology and Preparation of Process Qualification / Validation Report)
Post Exhibit / Validation batch documentation
Initiation of 'Post exhibit review checklist before commercial batch start-up' by Production

The SU R&D lead, scientist, or CRO scientist or lead is responsible for transferring the manufacturing process technology to the RU. The SU QC/Analytical Development Laboratory, R&D, or CRO, is responsible for transferring analytical methods and validating them at the receiving unit. R&D (if internal) communicate with the RU about the proposed transfer and provide a list of technical transfer documents (TTD). In another case, CRO will also communicate about the proposed transfer via email. Meetings and discussions with experts from the SU and RU, along with other stakeholders such as QC, production, and QA, will be organized by the R&D lead/scientist/CRO and the RU Manufacturing Science and Technology (MS&T) to discuss and finalize the product transfer. The RU Quality Assurance (QA) team initiate the local FDA license application procedure for a transferred product. After evaluating the technical transfer documents (TTD), the RU team will identify the prerequisites and requirements necessary for the technology transfer and delegate the necessary activities to the appropriate team members. These activities may include resource planning, procurement of raw and packing materials, equipment and material evaluations, environmental health and safety assessment, and compliance with regulatory requirements. The RU will also receive analytical method transfers from the SU QC lab/Analytical Development Laboratory or CRO and will procure any necessary new equipment, instruments, and testing materials for the proposed product.

These activities for execution include initiating a scale-up review checklist, manufacturing the batches according to approved procedures, testing them according to in-process controls and specifications provided by SU, evaluating the data and preparing a report, and assessing the need for further trial/scale-up batches if necessary. The process and parameters for exhibit batches are finalized based on the scale-up batches. The R&D (SU) team, along with the RU team, will evaluate the trial or scale-up data to create a list of critical quality attributes and rationale for the exhibit/submission batch. RU creates the Batch Manufacturing Record (BMR), Operational Packing Material List (OPML), Batch Packing Record (BPR), Process Validation/Qualification Protocol, and Quality Assessment (QRA). RU also initiate the change control process for the new product as per the change control SOP. Before the exhibit batch is manufactured, an MS&T team member distributes a "Checklist for exhibit batch start-up" to MS&T, production, packaging development, engineering, GSO, warehouse, EHS, QC, regulatory affairs, and QA departments for review and approval. The first batch of the development transfer product is manufactured in the presence of the R&D/CRO/Process Development/Production personnel. The exhibit/qualification/validation batches are manufactured per the approved Batch Manufacturing Record and Process Qualification/Validation Protocol, which have been prepared based on the approved Master Formula Card (MFC). RU QC shall conduct analytical testing and stability studies as outlined in the Technical Transfer Document (TTD) and per the approved stability protocol. The RU MS&T team creates the Process Qualification report/Process Validation report. The acceptance of the production process technology and conclusion on the product process Technology transfer from R&D (SU) to the manufacturing site (RU) is discussed in the Process Qualification/Validation report. Before the start of the first commercial batch, the production department circulate the "Post exhibit review checklist for commercial batch start-up" among the MS&T, production, packaging development, engineering, GSO, warehouse, EHS, QC, regulatory affairs, and QA departments for review and approval.

Site-to-Site Transfer

The site-to-site transfer (Fig. 4.7) process includes a transfer protocol that captures the experience from prior validations, annual product review documentation that includes information on process capability, rework history and procedures, process control trend summaries, deviations and investigations, and a strategy to establish equivalency between the sending unit (SU) and the receiving unit (RU).

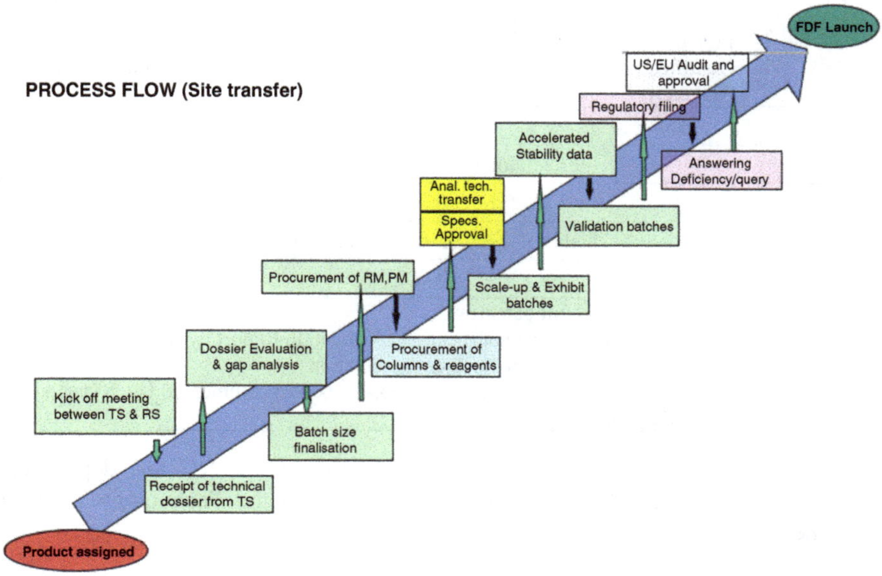

Fig. 4.7 Site-to-Site transfer

The technology transfer process for a drug product is a multiphase process that involves the acquisition of information, the planning of a transfer strategy, the execution of that strategy, and the comparison and evaluation of the results. The process is outlined in a sequential manner, but some activities may be done concurrently. There is also a general process flow chart that provides an overview of the technology transfer process from one manufacturing site to another:

Sharing of TT Documents by SU QA to RU QA as per TTD

⇓

Review of TT Documents by RU Team led by RU MS&T

⇓

Planning of Trial / Scale Up / Exhibit Batches at RU
(Indenting for RM / PM / Tooling / Change Parts, License Applications to FDA etc.)

⇓

Preparation of Technology Transfer Protocol (TTP) by RU
Initiation of scale-up review checklist by RU MS&T
Preparation and Issuance of documents for trial / scale-up batches (Sampling protocol, Specifications, STP, BMR, BPR, Cleaning method etc.)
Analytical Methods Transfer

⇓

Execution of Trial / Scale up batch(es)

⇓

Initiation of Checklist for exhibit batch start-up by RU MS&T Preparation and issuance of documents for exhibit batch(s) (BMR, Process Qualification / Validation Protocol, BPR, stability protocol, specifications, STP etc.)

⇓

Execution of exhibit batch(es)

⇓

Stability study of exhibit batch(es)

⇓

Evaluation of exhibit batch(es) Preparation of Process Qualification / Validation Report by RU Post Exhibit / Validation batch documentation Initiation of 'Post exhibit review check-list before commercial batch start-up' by Production

Upon confirmation of a technology transfer project, the RU QA department communicates with the SU QA department to request the Technology Transfer Document (TTD):

- License copy of the product(s), if required
- Master Formula Card (MFC)
- Material Safety Data Sheet (MSDS)
- Master Batch Manufacturing Record (MBMR)
- Operational Packaging Material List (OPML)
- Master Batch Packing Record (MBPR)
- Raw Material Specification (RMS) and Standard Test Procedure (STP)
- In-process Specification (IPS) and Standard Test Procedure (STP)
- Finished Product Specification (FPS) and Standard Test Procedure (STP)
- Packing Material Specification (PMS) and Standard Test Procedure (STP)
- Vendor Details and Vendor Qualification (VQ) Reports (API, RM and PM)
- Stability Testing Protocol
- List of Manufacturing and Packing Equipment and their Technical Specification
- Equipment Material of Construction (MOC) and compatibility assessments
- Tool Drawings (Punches, Blister Layouts, etc.)
- Stability Reports of Exhibit/Current commercial batches (Accelerated stability (ACC)/Long Term)
- Latest Annual Product Review (APR)
- Validation/Qualification Protocol for Cleaning Validation and test methods
- Process validation/Qualification protocol and reports
- Analytical method validation data (if required)
- Hold time study data (if required)
- Product dossier (if required)
- Specific SOPs related to process and cleaning
- Any other relevant documents

Table 4.1 Departmental responsibilities

Stages	Proposed responsible department
Stage 1a Formulation development	R&D
Stage 1a Analytical development	R&D
Stage 1a Process development	MS&T
Stage 1b Technology transfer/scale-up	MS&T
Stage 1b Analytical transfer	QC
Stage 2a Equipment/Utility/System qualification	Engineering
Stage 2b Process performance qualification	MS&T
Stage 3a Continued process verification	MS&T
Stage 3b Continued process verification	MS&T

The above list may vary depending on the transferred product and will be considered case-by-case. The SU department should provide the latest approved versions of the above-mentioned documents. The SU QA department should share all documents in the TTD with the RU QA department, preferably using an FTP server, with proper folder and subfolder naming. The SU QA department initiates and fills out the Technology Transfer Document (TTD) Checklist and communicates with the RU QA department regarding the sharing/uploading of documents, including the path/location and folder contents on the FTP server. After receiving the TTD checklist, the RU QA department checks the availability of the documents shared against the TTD checklist and records its receipt. This should also be checked by the RU MS&T team members and approved by the RU QA department. The RU MS&T team member should organize meetings/discussions with the SU experts in case of any queries, clarifications, and/or information required for understanding.

The planning phase of a technology transfer strategy involves several steps to ensure the successful transfer of technology (Table 4.1). These steps include initiating the local license application procedure, evaluating the technology to be transferred, identifying prerequisites and requirements for the transfer, preparing a Technology Transfer Protocol (TTP) that lists the intended stages of the transfer, and assigning a unique identification number to the TTP. Additionally, the planning phase may include a feasibility study of the facility and equipment at the RU and the SU to ensure that the equipment and facilities are similar and operate according to similar principles. The impact of manufacturing new site transfer products on the products currently manufactured at RU with the same equipment should be determined. A comprehensive technical gap analysis between SU and RU, including technical assessment and potential regulatory gaps assessment, especially concerning potential post-approval changes, should be performed as needed. Regulatory requirements/local FDA license applications, product licenses, and excise formalities are performed. Acquiring samples for analytical method transfer (AMT), reference, testing, and/or comparison, and preparation of master documents, specifications, test methods, MFC, BMR, BPR, OPML, and stability protocol is critical. Indenting the RM, PM, machine change parts, new machines, instruments, chemicals, reagents, columns, and standards for testing the proposed product happens at this stage. The modifications/changes/new installation, EHS (Environment Health and Safety) requirement and preparation of specific safety procedures, Material Safety Data Sheet (MSDS), if necessary, for the TT and/or manufacturing of the proposed product at the RU is conducted. The site prepares the necessary qualification and validation documents like process validation/qualification protocol, cleaning validation protocol, and AMT protocol. The TTP is reviewed by the RU team member and approved by the RU and SU MS&T Head, RU Manufacturing Head, and RU Quality Assurance Head. Similarly, the AMT protocol must be prepared and approved for execution by the SU/RU QC.

In the execution phase, the qualification and validation, as per the respective qualification/validation protocol and AMT as per the approved AMT protocol, are performed. Before manufacturing the trial/scale-up batch, the scale-up review checklist is initiated by RU MS&T. Then the review and necessary comments from the concerned department are received before execution of the trial/scale-up batch. The trial/scale-up batch(s) are manufactured per the approved MFC and monitored by SU members wherever required. The RU MS&T team prepares the scale-up report after the completion of trial/scale-up batches. If there are any changes after the completion of the scale-up, the SU and RU MS&T teams assess/evaluate the requirement for further trial/scale-up batches. Before the manufacturing of the exhibit batch, RU MS&T or the Packaging Development team circulate the "Checklist for first exhibit batch start-up" among Production, Packaging Development, QC, EHS, Warehouse, GSO, Regulatory Affairs, and QA departments for review and approval. Manufacturing of exhibit/validation batch (es) follows the approved batch manufacturing record and process qualification/validation protocol prepared based on approved MFC. Packaging of the batch(s) is performed per the approved batch packaging record prepared based on approved OPML. Process qualification/validation report preparation, completing the analytical testing, and conducting the stability study as per the approved stability protocol are the other activities to complete.

The post-assessment phase (comparison and evaluation of results) is initiated with the RU MS&T team preparing the technology transfer report (TTR) after the successful technology transfer. Results and/or reports of all activities conducted in the execution are compiled, analyzed thoroughly, and compared with the corresponding results of the SU. TTR is a documented summary of a specific technology transfer project, listing procedures, acceptance criteria, results achieved, and conclusion. The differences, changes, noncompliance, or deviations observed are discussed and addressed. The TTR is identified with a unique identification numbering. In case of any changes implemented at the SU during/after the drug product transfer, it is communicated to the RU in writing. SU QA communicates the change control to RU QA. The TTR shall be reviewed by RU team members and approved by the RU and SU MS&T Head, RU Manufacturing Head, and RU Quality Assurance Head. Approval of the TTR concludes the product TT activity. Before the start of the first commercial batch, the production department circulates the "Post exhibit review checklist for commercial batch start-up" to MS&T, production, packaging development, engineering, GSO, warehouse, EHS, QC, regulatory affairs, and QA department for review and approval.

References

1. PDA (2022). PDA technical report no. 65 revised 2022 (TR 65) technology transfer. Parenteral Drug Association
2. WHO (2021) Guidelines on technology transfer in pharmaceutical manufacturing. World Health Organization
3. Pattanaik A (2014) The complexity of project management in the pharmaceutical industry, Paper presented at PMI Global Congress
4. FDA (2022) 21 CFR part 210, 211: current good manufacturing practice in manufacturing processing, packing, or holding of drugs. FDA

Chapter 5
Scaling-Up of Solid Orals: Granulation, Drying, Size Reduction, Blending, Compression, and Coating Technologies

Abstract The scaling-up is a crucial step in the development and manufacturing process. The goal is to optimize the process to ensure consistent and efficient production of the desired dosage form. One aspect of this process is the equipment classification, governed by the FDA's SUPAC (Scale-Up and Post-Approval Changes) guidelines. These guidelines provide a framework for ensuring that any changes made during the scaling-up process do not adversely affect product quality, safety, or efficacy. Scale-up factor calculation methods are an important aspect of the process, as they help determine the appropriate scaling factor required to produce larger batches of the product. These methods take into account various factors such as equipment capacity, mixing time, and other critical process parameters. Manufacturing technologies such as granulation, drying, size reduction, blending, compression, and coating are essential steps in the production of solid oral products. Granulation is the process of forming granules from powder particles, which improves the flowability and compressibility of the product. Drying removes moisture from the granules, and size reduction is employed to reduce particle size for improved solubility and bioavailability. Blending ensures that the active ingredients and excipients are uniformly mixed, and compression is used to produce tablets of the desired shape and size. Coating provides a protective layer to the tablet and can improve the appearance, stability, and taste of the product. Scaling-up requires a thorough understanding of equipment classification, scale-up factor calculation, and the manufacturing technologies.

Keywords Scale-Up · Solid Orals · Stage 1b · Manufacturing Technologies · Scale-Up Factor

Equipment Classification: SUPAC

SUPAC stands for Scale-Up and Post-Approval Changes, and it is a guideline issued by the U.S. Food and Drug Administration (FDA) [1]. The guideline provides information on the equipment classification and the process changes that can be made to a drug product after the FDA has approved it. According to SUPAC, equipment is classified into three categories: Category I, II, and III. Category I

A. B. Pazhayattil et al., *Technology Transfer*, AAPS Introductions in the Pharmaceutical Sciences 10, https://doi.org/10.1007/978-3-031-32192-4_5

equipment changes are considered minor and do not affect the critical quality attributes of the drug product. Examples of Category I equipment changes include minor changes to the equipment design, the use of different construction materials, or the addition of new equipment components. These changes do not require FDA review and can be implemented without prior notification to the FDA. Category II equipment changes are considered moderate and may affect the critical quality attributes of the drug product. Examples of Category II equipment changes include changes to the equipment design that may impact the critical quality attributes of the drug product, the use of different materials of construction that may affect the critical quality attributes of the drug product or the addition of new equipment components that may impact the critical quality attributes of the drug product. These changes require prior notification to the FDA but do not require FDA review or approval. Category III equipment changes are considered major and are likely to affect the critical quality attributes of the drug product. Examples of Category III equipment changes include changes to the equipment design that is likely to impact the critical quality attributes of the drug product, the use of different materials of construction that are likely to affect the critical quality attributes of the drug product or the addition of new equipment components that are likely to impact the critical quality attributes of the drug product. These changes require prior notification to the FDA and FDA review and approval. The classification of equipment changes is important for companies to ensure that their changes comply with FDA regulations and do not negatively impact the safety and efficacy of the drug product.

Scale-Up Basic Approach

Scale-up basic approach revolves around:

- Geometrical similarity.
- Kinematic similarity.
- Dynamic similarity.

If the above three criteria are satisfied, the scaling-up of parameters should be linear. In the pharmaceutical industry "Scale-up" refers to the processes that are needed for successful transitions from drug discovery to product development to clinical trials to full-scale commercialization. Scale-up is an inevitable part of the product lifecycle of every successful drug, and each time it is required, a meticulous process must be followed to ensure that the end result is identical to the product formulation as originally devised. Scale-up of any pharmaceutical manufacturing process entails a skillful combination of art, experience, science, and engineering. Application of statistical methods is constantly increasing to help with design of experiments and development of empirical relationships between process parameters, material attributes, and quality attributes.

The scale-up approach has been used in physical sciences, such as fluid dynamics and chemical engineering, for a long time. This approach is based on process similarities between different scales and employs dimensional analysis that was developed a century ago and has gained applications in many industries, especially in chemical engineering. Dimensional analysis is a method for producing dimensionless numbers that completely characterize the process parameters. The analysis can be applied even when the equations governing the process are not known. According to the theory of models, two processes may be considered completely similar if they take place in similar geometrical space and if all the dimensionless numbers necessary to describe the process have the same numerical value. The scale-up procedure is simple if scale-up is executed in similar geometrical and dimensional environment. Dimensionless numbers, such as Reynolds and Froude numbers, are frequently used to describe mixing processes. Scientists are routinely concerned with problems of water-air or fluid mixing in vessels equipped with turbine stirrers in which scale-up factors can be up to 1:70. However, scale-up challenges arise when similarity is not possible for geometrical and dimensional parameters. One way to eliminate potential scale-up challenges is to develop formulations that are very robust with respect to processing conditions. A comprehensive database of excipients detailing their material properties may be indispensable for this purpose. In practical terms, this cannot be achieved without some means of testing in a production environment, and, since the initial drug substance is usually available only in small quantities, some form of simulation is required on a small scale. In a very complex manufacturing process scale-up, it may need minor modifications in existing process flow, addition or deletion of extra processing steps, minor changes in composition, changes in equipment principles, etc.

Any change in a process of making a pharmaceutical dosage form is a regulatory concern. Scale-Up and Post Approval Changes (SUPAC) are of special interest to the FDA, as is evidenced by a growing number of regulatory documents released in the past several years by the Center for Drug Evaluation and Research (CDER), including Immediate Release Solid Oral Dosage Forms (SUPAC-IR), Modified Release Solid Oral Dosage Forms (SUPAC-MR), and Semisolid Dosage Forms (SUPAC-SS). Additional SUPAC guidance documents that are being developed include Transdermal Delivery Systems (SUPAC-TDS), Bulk Actives (BACPAC), and Sterile Aqueous Solutions (PAC-SAS). Collaborations between the FDA, the pharmaceutical industry, and academia in this and other areas have recently been launched under the framework of the Product Quality Research Institute (PQRI). Scale-up problems may require post-approval changes that affect formulation composition, site, and manufacturing process or equipment (from the regulatory standpoint, scale-up, and scale-down are treated with the same degree of scrutiny). In a typical drug development cycle, once a set of clinical studies has been completed or a NDA/ANDA has been approved, it becomes very difficult to change the product or the process to accommodate specific production needs. Such needs may include changes in batch size and manufacturing equipment or process. Post-approval changes in the size of a batch from the pilot scale to larger or smaller production scales call for submission of additional information in the application, with a specific requirement that the new batches are to be produced using similar test equipment and in full compliance with CGMPs and the existing SOPs. Manufacturing changes may require new stability, dissolution, and in vivo bioequivalence testing. This is especially true for Level 2 equipment changes (change in equipment to a different design and different operating principles) and the process changes of Level 2 (e.g., in mixing times and operating speeds within application/validation ranges) and Level 3 (change in the type of process used in the manufacture of the product, such as from wet granulation to direct compression of dry powder). Scale-up should be done after getting all information from R&D. It involves transfer of technology as well as transfer of knowledge. Based on various regulatory agency requirements in the pharma industry, the ideal scale-up can be up to 10 times of pilot scale; however, based on suitable supplement, scale-up beyond 10 times of pilot scale may be approved by regulatory authorities. In an ideal scenario based on the targeted market, the final maximum commercial volume in units can be finalized at the initial stage of product selection for development. This helps the organization to finalize the finale commercial batch size and hence subsequent pilot batch size. An early-stage decision on the scale of the product preparation will help to finalize development strategy and effective utilization of available commercial resources. During development of any product, if the primary scale-up feasibility is taken care of, the subsequent challenges during actual execution of scale-up can be reduced. However, scale-up is not always fully predictable, hence it is necessary to go through a systematic approach in this regard. The development of robust formulation and process using design of experiments (DoE) as well as a good understanding the critical v/s noncritical parameters for each unit operation are major determining factors for success v/s failure on scale-up. A successful drug product may go through a scaling process several times during its lifecycle. The laboratory scale batches that expand to pilot-scale and finally to commercial scale production may be just a single iteration in a product's evolution. Popular products may also expand production to other manufacturing facilities or even other countries. Conversely, when demand inevitably begins to shrink, a similar process will be used in scaling down production to appropriate levels. Parameters that require adjustment based on changes in scale to maintain consistent processing conditions are known as scale-dependent parameters, and they are typically calculated using dimensionless numbers through scale-up calculators. For example, in a drying operation utilizing a fluid bed dryer, the air flow required for fluidization is a scale-dependent parameter that must be adjusted with each scale change. In contrast, parameters that remain consistent regardless of scale are known as scale-independent parameters. For instance, in the same drying operation utilizing a fluid bed dryer, the inlet air temperature would remain constant regardless of the scale of the operation. To achieve successful scale-up, several parameters should be taken into consideration. These include flexibility, cost, dependability, innovation, and product quality.

In addition, it is essential to acknowledge that effective communication plays a critical role in the successful transfer of formulations and processes. Scale-up is generally defined as the process of increasing batch size and procedure by applying the same process to different output volumes or migrating a process from the lab scale to the pilot plant scale or commercial scale. Objectives of scale-up process include:

- To meet predefined CQA parameters.
- To understand what makes these processes similar, to identify and to eliminate many scale-up problems before investing large sums of money on a production unit.
- To maintain efficacy through changes in processes or composition which can help to overcome scale-up issues.
- To closely examine the formula to determine its ability to withstand large scale and process modifications.
- To review a range of relevant processing equipment in order to determine which would be most compatible with the formulation as well as the most economical, simple, and reliable in producing the product.
- To assess production rates and future market requirements.
- Difficulties faced during scale-up.
- Equipment geometry may not be similar across scale.
- Equipment make and model may be different.
- Equipment capability to measure some critical parameters may differ across scale.
- Lack of established correlation to scale-up for critical process parameters.
- Interpretation of powder characterization is not extensive.
- Lack of automation may lead to manual errors.
- Lack of process analytical tools to determine end point and control process parameters in-line to achieve desired quality attributes.

Batch Size Calculation

Determining the appropriate batch size is a critical decision in the pharmaceutical industry that is influenced by several factors, such as the type of manufacturing technology used, regulatory requirements, supply chain needs, and operational planning considerations. To properly understand batch size, it is important to distinguish between various manufacturing methods, including batch, continuous, semi-batch, and semi-continuous. Batch manufacturing involves loading all materials at the beginning of the process and unloading them at the end, as seen in bin blending and lyophilization. Continuous manufacturing, on the other hand, involves simultaneous loading and unloading of materials throughout the process and is commonly used in industries such as petroleum refining, food processing, and increasingly, pharmaceutical manufacturing. Semi-batch, or fed-batch, manufacturing is used in wet granulation, tablet coating, and fermentation, where materials are added during the process and discharged at the end. In semi-continuous manufacturing, materials are loaded and unloaded simultaneously, but only for a defined period, as seen in roller compaction, tablet compression, and encapsulation. In semi-continuous manufacturing, the process output is not dependent on batch size if the material input is configured to produce consistent output, making a fixed batch size unnecessary.

Granulation Technologies

High Shear Granulation (RMG-Rapid Mixer Granulator)

High shear granulation using RMG-Rapid Mixer Granulator involves factors such as densification, agglomeration, shearing and compressing action of the impeller, mixing, granulation and wet massing, possibility of over-granulation due to excessive wetting, possibility of producing low porosity granules, liquid bridges, coalescence, breakage of the bonds, specific surface area, moisture content, intragranular porosity, heating, evaporation, and mean granule size.

Granule Growth Mechanism

There are two mechanisms for describing granule formation and growth, which are nucleation of particles and coalescence between agglomerates. Granulation growth occurs in three stages: granule nucleation and binder distribution, granule consolidation and growth, and granule attrition and breakage. In formulations driven by nucleation and subsequent growth induction, granule properties depend on nucleation and wettability, so the focus is on the nucleation zone, also known as the wetting or spray zone. This is the area where liquid and solid first come into contact and form loose agglomerates. Within this zone,

$$\text{Nuclei formation} = \int \text{kinetics and wetting thermodynamics}$$

$$\text{Binder Dispersion} = \int \text{effective mixing of powder and binder}$$

Nucleation

Nucleation is the combination of single drop behavior and multiple drop interactions. The single drop behavior is controlled by Tp or drop penetration time, which is determined by material properties. On the other hand, the dimensionless spray flux or Ψa is controlled by operating variables in the spray zone. Tp controls the flux of drops on the surface of powder particles, and if there are too many drops, coalescence may occur, leading to broad size distribution.

$$\Psi\text{a} = 3V/2A\text{Dd}$$

where V is the volumetric spray rate, A is the area flux of powder traversing the spray zone, and Dd is the droplet diameter.

Coalescence

Coalescence occurs when two agglomerates collide and undergo plastic deformation due to surface water. The outcome of coalescence is dependent on the absolute moisture content versus the liquid saturation of the granules.

$$S = H(1 - \varepsilon)\rho/\varepsilon$$

where H is the moisture content on a dry basis, ε is the granular porosity, and ρ is the particle density of the feed material.

In high-shear mixers, the rapid movement of the mixer blades causes the powder material to enter a state of spiral fluidization at the base of the mixing bowl. This creates a vortex effect that enables faster and more even distribution of the ingredients. The bowl is shaped to facilitate an optimal and consistent flow of the processing material. The intense mixing action ensures that active substances are distributed homogeneously within a few minutes. The binder is added through a charging hole at the top cover, which uniformly moistens and aggregates the particles into large granules. Finally, the vertically rotating, specially designed chopper blades shear the circulating powder and break down lumps and aggregates (Fig. 5.1).

Fig. 5.1 Granule growth in high-shear mixer

Effect of Feed Material on Granule Growth in High Shear Mixer

Granule growth in high shear mixers is highly dependent on the amount of liquid phase and processing conditions such as impeller rotation speed and processing time.

The following criteria are considered for process selection in granulation:

- High drug load in the formulation, typically ranging from 60% to 80% by weight of the total blend.
- Poor compactability of the active pharmaceutical ingredient (API) as a direct compression blend or a tendency to stick to metal surfaces in the dry state.
- Low dose number (<10), indicating a relatively high solubility of the drug in biologically relevant media.
- Granulation facilitates intimate contact of the drug with wetting and stabilizing agents, thereby improving the solubility and stability of the drug.
- The innovator product employs the same granulation process.
- As the API load in the formulation increases, certain material attributes of the API become important, such as the particle size and distribution, particle morphology and associated wettability, and the impact of applied shear on form change. Additionally, the impact of moisture and heat on chemical degradation is also a significant consideration.

Granulation properties can be influenced by various factors including equipment variables such as the type and design of the granulator, process variables such as the speed, time, and temperature of granulation, and product variables such as the particle size, shape, and moisture content of the raw materials. The equipment variables play a significant role in determining the shear forces and hence the granulation properties. The design of the bowl, impeller (Fig. 5.2), and chopper can impact the mixing efficiency, granule size distribution, and granule strength. Shear forces in a high shear mixer are very dependent on the mixer construction. Consequently, apparatus variables are more essential when using high shear mixers. Size and shape of the mixing chamber, impeller, and chopper differ in different high shear mixers. A small change in shape, size, or inclination of the blade tips has a significant effect on the impact of the mass. Bowl specification is a crucial aspect of design, which includes dimensions such as height, width, effective volume, and height to width ratio. To optimize and scale-up the design, the percentage occupancy or fill volume of the bowl with the powder mass needs to be considered. This allows for normalization of any variations in bowl geometry, such as the height to width ratio, that may be observed across different equipment and scales. When it comes to impeller design, various factors such as attack angle, blade design, swept volume, and whether the impeller is top or bottom mounted into the bowl need to be taken into account. The PMA impeller is the standard design used for conical bowls. However, the M8 impeller has an innovative swept-back design, which offers improved mixing characteristics, faster processing of the blend to granule, and a better-defined end point.

Fig. 5.2 Impeller

Table 5.1 Relative swept volume

Equipment	Inclination angle of impeller	Relative swept volume/s impeller speed
Diosna P 25	36°	1.37
PMA 25	29°	0.75
PMA 65	30°	0.71
Diosna P 50	39°	1.08
PMA 150	30°	0.61
Diosna P 250	54°	0.52

The relative swept volume (Table 5.1) is a parameter that is used to quantify the amount of work input on the material in a mixer. It is defined as the volume swept out per second by the impeller divided by the volume of the mixer. This parameter is important because it provides a way to relate the work input to the densification of the wet mass in the mixer. In general, a higher relative swept volume indicates that more work is being done on the material, which can lead to greater densification of the wet mass. However, the relationship between the relative swept volume and densification is not always straightforward, as other factors such as impeller design, mixing time, and shear rate also play a role. Therefore, it is important to consider the relative swept volume in conjunction with other parameters when optimizing the mixing process. The relative swept volume seems to be an appropriate parameter when comparing the effect of size and construction of the mixing tools.

When it comes to chopper design, there are two options to consider. The first is the standard U-shaped chopper that is designed for use in a conical bowl. The second option is a flush-mounted multiblade chopper, which is specifically designed to improve the dispersion of the binder solution into the powder mass.

The design of the spray nozzle (Fig. 5.3) and its positioning on the lid of the bowl are critical factors that affect the delivery and dispersion of the binder solution to the powder mass during the mixing process. It is important to carefully consider the optimal location of the spray nozzle to avoid any localized over-wetting of the powder mass. In addition to the spray nozzle design, the choice of pump design is also crucial in achieving precise and consistent delivery of the binder solution.

Fig. 5.3 Spray nozzle

Integrated peristaltic pumps are particularly well-suited for this purpose. When selecting a pump, the capacity of the pump and the diameter and flexibility of the tubing used to deliver the binder solution are key considerations. By considering these factors, it is possible to ensure that the binder solution is delivered in a controlled and uniform manner, leading to a more homogenous mixture and improved product quality.

There are six important granulation process variables: impeller and chopper rotation speed, mixer load, liquid addition method, liquid flow rate, and wet massing time. During the granulation process, it is important to follow several key steps, including keeping the chopper off during dry mixing, selecting appropriate impeller RPM, adding binder through a nozzle or sprinkler in a single step, keeping the chopper on during binder addition, fixing wet massing time, using peristaltic pumps for consistent flow rate, and avoiding excessive stirring of the binder. It is also recommended to prepare a known excess of the binder to improve accuracy of addition and to consider a larger pump head and/or wider tubing diameter for binders of high viscosity before using positive displacement pumps. Product variables that must be considered in granulation include characteristics of the feed materials such as particle size, solubility in the liquid binder, wettability, and packing properties. The amount and characteristics of the liquid binder, including surface tension and viscosity, are also important. Formulation-related variables include approximate matching of particle size, addition of the binder as a dry powder or dissolved in an aqueous solution, and the addition of microcrystalline cellulose. It is important to avoid using anhydrous lactose or similar excipients that would undergo a form change during wet granulation and to always add the lubricant during the final blending step.

Granulation Endpoint

Granulation end point (Fig. 5.4) refers to the stage at which the granulation process is stopped, and the resulting granules meet the desired product specifications. The end

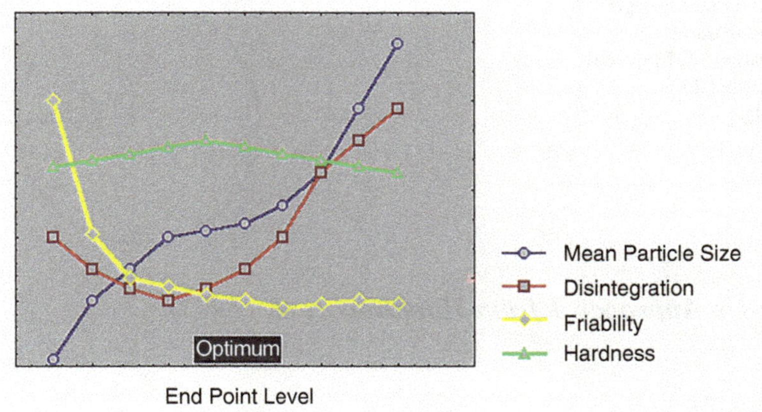

Fig. 5.4 Granulation end point

Table 5.2 Scale-up prerequisites

Granulator size	4 l	10 l	75 l
Granulation batch size (kg)	0.6	1	9.72
Diameter of the granulator bowl (cm)	17	24.6	52.5
Fill height/Diameter	0.27	0.15	0.15

point is determined by several factors (Table 5.2), including the target particle size mean, particle size distribution, granule viscosity, and granule density. However, determining the end point can be challenging, especially when encountering changes in equipment size and model, which can affect the reproducibility of the endpoint. There are several methods for determining the granulation end point, including hand testing, emerging methods such as acoustic emission and image processing, offline methods like torque rheology and particle size analysis, and in-line instrumentation like main impeller motor amperage, power, and shaft torque. However, hand testing is a qualitative, subjective, and inconsistent method that relies on the operator's experience and judgment. Emerging methods and off-line methods are more objective and reliable but require additional equipment and analysis time. In-line instrumentation provides real-time feedback and is useful for monitoring the process and adjusting in real time.

Ratio of fill height to granulator diameter at all scales should be kept constant. The spraying time and the dimensionless spray flux need to be kept constant.

4 liters RMG – $\Psi = 0.11$. For constant spraying time, the Ψ values for larger scales with single nozzle are calculated assuming 100 μm drop size:

10 liters RMG – $\Psi = 0.15$ and 0.18 (Close to small scale)

75 liters RMG – $\Psi = 0.34$ and 0.50 (Too high compared to small scale, therefore higher amounts of lumps should be expected. Can be lowered by using two nozzles)

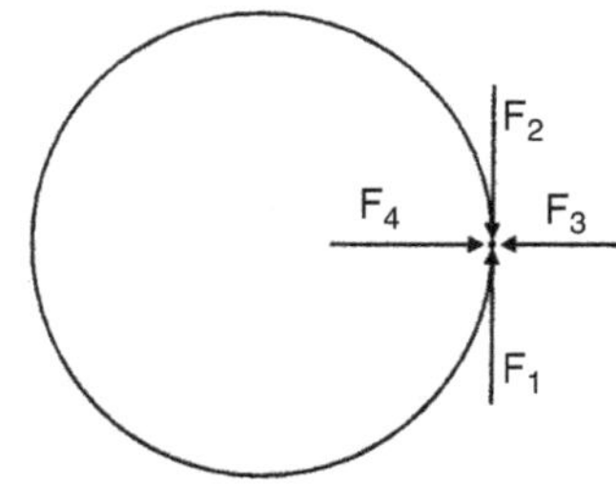

Fig. 5.5 Forces in high shear granulation Acceleration F1, Frictional F2, Centripetal F3, Centrifugal F4

Scale-Up Approach 1 from Horsthius et al. (1993)

$$\mathrm{Fr} = n^2 d/g$$

where n is impeller speed, d is impeller diameter, and g is gravitational constant.

They concluded that maintaining an equal Froude's number at different scales resulted in comparable particle size distribution.

The Froude number is utilized to compare mixers based on certain criteria. As a dimensionless quantity, it is not reliant on the size of the machine. It represents the ratio of centrifugal force to gravitational force and can serve as a measure of dynamic similarity among mixers (Fig. 5.5).

Scale-Up Approach 2 from Rekhi et al. (1996)

The impeller tip speed should be kept constant throughout the scale-up process. Granulating liquid volume proportional to batch size: The amount of granulating liquid used should be proportional to the batch size. The wet massing time should be inversely proportional to the RPM of the mixer. This means that as the RPM increases, the wet massing time should decrease to maintain consistent mixing intensity and granule properties.

RPM to be scaled down and scaled up according to Tip speed:

$$2^* \left(\prod\right)^* R^* N$$

Scale-Up Approach 3, Using Power Number Correlations Landin, M., P. York, M.J. Cliff (1996)

Based on the concept of similarity, there are three types of similarities that can be observed: geometric similarity, kinematic similarity, and dynamic similarity. To measure the similarity between different systems or objects, dimensionless numbers are used. Three commonly used dimensionless numbers are Newton (power) number (Ne), Froude number (Fr), and Reynolds number (Re).

The Newton (power) number (Ne) represents the ratio of power consumption to the density and size of particles being mixed. It can be calculated using the formula:

$$\mathrm{Ne} = P / \left(\rho n^3 d^5\right)$$

where P is the power consumption, ρ is the specific density of particles, n is the impeller speed, and d is the impeller diameter.

The Froude number (Fr) represents the ratio of inertial forces to gravitational forces and can be calculated using the formula:

$$\mathrm{Fr} = n^2 d / g$$

where n is the impeller speed, d is the impeller diameter, and g is the gravitational constant.

The Reynolds number (Re) represents the ratio of inertial forces to viscous forces and can be calculated using the formula:

$$\mathrm{Re} = d^2 n \rho / \eta$$

where d is the impeller diameter, n is the impeller speed, ρ is the fluid density, and η is the the dynamic viscosity.

The power number relationship is given by the formula:

$$\mathrm{Ne} = K(\mathrm{Re} \cdot \mathrm{Fr} \cdot h/D)n$$

where K is a constant, h is the height of the powder bed, and D is the diameter. This relationship can be applied to a series of geometrically similar high shear mixers, regardless of their scale, as it is dimensionless.

The experimental procedure consists of several sequential steps. Initially, the powders are charged into the mixer, followed by switching on the mixer, and recording the power reading. Water is then added at a constant rate to the mixer. At certain predetermined water contents, the power reading of the mixer is noted, and a sample is taken. The density of the sample is measured, and the viscosity is determined. Using the obtained values, the power, Reynolds, and Froude numbers are calculated. Finally, the power number relationship is plotted based on the calculated numbers. The experimental s aims to analyze the effects of mixing conditions on the granulation process.

Drying Technologies

In a direct heating static solid bed dryer, the solid material is placed in a stationary bed and heated directly by a heat source, such as hot air or gas. The subclasses of this type of dryer include tray and truck dryers, where the solid material is placed on trays or trucks and moved into the dryer, and belt dryers, where the material is placed on a conveyor belt and moved through the dryer. In a direct heating moving solid bed dryer, the solid material is placed on a moving bed and heated directly by a heat source, such as hot air or gas. The subclasses of this type of dryer include rotating tray dryers, where the solid material is placed on trays that rotate and move through the dryer, and horizontal vibrating conveyor dryers, where the material is placed on a conveyor that vibrates horizontally and moves through the dryer.

In a direct heating fluidized solid bed (FBD) dryer, the solid material is placed in a bed and heated directly by a heat source, such as hot air or gas. The bed is fluidized by introducing air or gas, which causes the solid particles to behave like a fluid and move freely, allowing for more efficient drying. In a direct heating dilute solid bed (spray drier/flash drier) dryer, the solid material is first converted into a liquid or slurry form and then atomized into fine droplets, which are sprayed into a hot gas stream. The droplets get dried instantly due to the intense heat and high gas velocity, resulting in a dry powder product. Spray dryers are typically used for drying liquids and pastes, while flash dryers are typically used for drying wet solids.

In a direct conduction heating, moving solid bed dryer, the solid material is placed on a moving bed and heated directly by conduction through contact with a heated surface, such as a heated drum or wall. The subclasses of this type of dryer include a paddle dryer, a rotary (tumble) dryer, and an agitation dryer. A paddle dryer is where the solid material is placed on a moving bed that is agitated by paddles. A rotary (tumble) dryer is where the solid material is placed in a rotating drum that tumbles the material as it is heated. Agitation dryer is where the solid material is placed in a container that is agitated as it is heated. Each subclass uses different mechanical means for moving the solid bed, but all use direct conduction heating to dry the material.

In a direct conduction heating, static solid bed dryer, the solid material is placed in a stationary bed and heated directly by conduction through contact with a heated surface, such as a heated plate or wall. The material remains in a static position as it is dried. In direct conduction lyophilization, also known as freeze-drying, the solid material is frozen and then placed in a vacuum chamber. The frozen material is then heated by conduction through contact with a heated surface, such as a heated shelf. As the material is heated, the frozen water within it sublimates (changes from a solid to a gas), leaving behind a dry solid product. In direct radiant heating (microwave dryer), the solid material is exposed to microwave radiation, which heats the material directly by causing the water molecules within it to vibrate and generate heat. The material remains in a static position as it is dried. Microwave dryers are typically used for wet solids but can also be used for liquids and pastes.

Granulation, Drying, e.g., Fluid-Bed Processors

Fluid bed processors (Fig. 5.6) process bulk materials in powder or granular form. They use a fluidizing air stream to suspend the material, which increases the surface area that interacts with the process area. This improves heat and mass transfer, allowing for efficient heating, cooling, drying, and moisturizing/wetting of the material. The equipment can also be used to build particle size by spraying binders over the suspended powder. The liquid can spray the powder in various ways, such as top, bottom, or tangential spraying.

Several factors can influence granulation in fluid bed processors (Tables 5.3 and 5.4), including:

Raw materials: The properties of the starting materials, such as particle size, shape, and moisture content, can affect the granulation process.

Binder: The type and concentration of the binder used can affect the strength and cohesion of the granules.

Nature of solvent: The choice of an aqueous or nonaqueous solvent can affect the granulation process, as well as the properties of the final granules.

Machine parameters: Various machine settings can also affect granulation, including fluidization air flow, inlet air temperature and humidity, product temperature, spray rate, and atomization air pressure.

Fig. 5.6 Fluid bed processors

Table 5.3 Variables

Material variables	Scale dependent/Independent
Hygroscopic nature of drug	Scale independent
Equipment Variables	
Spray nozzle diameter	Scale independent
Spray gun position	Scale dependent
Filter bag type	Scale independent
Dutch mesh diameter	Scale independent
Dispersion/Solution preparation variables	
Selection of stirrer	Scale dependent
Viscosity	Scale independent
Delivery pump	Scale dependent
ID/OD of silicon tubing	Scale dependent
Holding time	Scale dependent
Prewarming variables	
Inlet temperature	Scale independent
Product temperature	Scale independent
Fluidization air flow (CFM)	Scale dependent
Inlet air humidity	Scale independent
Prewarming time	Scale dependent
Filter shaking interval/pause	Scale independent
Spraying variables	
Occupancy (batch size)	Scale independent
Inlet temperature	Scale independent
Product temperature	Scale independent
Exhaust temperature	Scale independent
Fluidization air flow (CFM)	Scale dependent
Atomization air pressure	Scale dependent
Atomization air volume	Scale dependent
Spray rate	Scale dependent
Inlet air humidity	Scale independent
Absolute humidity	Scale independent
Dew point	Scale independent
Filter shaking interval/Pause	Scale independent
Drying variables	
Inlet temperature	Scale independent
Exhaust temperature	Scale independent
Fluidization air flow (CFM)	Scale dependent
Inlet air humidity	Scale independent
Drying time	Scale independent

Table 5.4 Controls

Stage	Process parameter	Responses to be studied		
		Physical	Chemical	PAT
Top spray granulation	Inlet temperature Product temperature Spray rate Atomization air pressure Fluidization air CFM Inlet air humidity	Description PSD Yield	Assay CU LOD RS	NIR Mass spectra
Drying	Inlet temperature Product temperature Spray rate Atomization air pressure Fluidization air CFM Inlet air humidity	Description PSD Yield	Assay LOD RS	NIR Mass spectra

Table 5.5 Batch size details

Parameter	CASE- A		
	Batch size	Scale	FBE capacity
Batch size of exhibit batch	141,000 units/162.60 kg	1×	250 L
Trial/pre PPQ batch	380,000 units/438.25 kg	2.7×	800 L

FBP- Scale-Up Calculation

If the fill ratio remains the same in all scales, then in any scale, the air velocity should remain the same: Air velocity (Feet/Min) = Airflow (Cubic Feet/Min)/Base plate area (ft^2)

$$\text{Spray rate of Scale B Batch } (\text{g/min}) \ (\text{Q2}) = \text{Q1} \times \text{A2}/\text{A1}$$

[Q1- Spray rate of Scale A Batch (g/Min), A2- Base Plate Area of Scale 2, A1- Base Plate Area of Scale 1. Differential pressure across the bowl should be maintained].

Scale Up/Down of Batch Size

Fixing the batch size is a preliminary step in scale-up; batch size is calculated on the higher scale by keeping % occupancy constant. For fixing the batch size, the working volume of the equipment should be taken into consideration. The following equation can be used to calculate the batch size (Tables 5.5, 5.6, 5.7, and 5.8):

$$B = V \times D$$

[Where B = Batch size in kg, V = working vol. of the container, D = product bulk density (g/cc)].

Table 5.6 Scale-up factor

	CASE- A	
Parameter	FBE base plate diameter (m)	FBE base plate area (mm^2)
Exhibit batch	0.727	0.4149
Trial/pre PPQ batch	1.005	0.793
Scale-up factor formula	Scale-up factor = (base plate area of large scale/Base plate area of small scale)	
Calculated scale-up factor	0.727/0.4149 = 1.911	

Size Reduction/Separation Technologies

Size reduction is a mechanical process that reduces the size of particles or aggregates by applying mechanical stress to the material to be crushed or ground. Depending on the material's properties, this stress can cause the material to yield, fracture, or exhibit a combination of both elastic and viscous properties. During milling, particles can be subjected to different forces, such as shear, compression, impact, or tension. These forces work to size by breaking them down at their weakest points, including the binder-particle interface, flaws in the individual ingredient particles, or a combination of these. Size reduction is an important process for creating uniform and consistent particles. The choice of equipment and methods used for these processes can have a significant impact on the properties of the final product, and it is important to select the appropriate equipment and methods to achieve the desired outcome.

The Fluid Energy Mills subclass has no moving parts. It is primarily distinguished by the configuration and shape of their chambers, nozzles, and classifiers, including tangential jet, loop/oval, opposed jet, opposed jet with dynamic classifier, fluidized bed, fixed target, moving target, and high-pressure homogenizer. Impact Mills, a subclass, are primarily distinguished by the configuration of the grinding heads, chamber grinding liners (if any), and classifiers, including hammer air swept, hammer conventional, and pin/disc and cage. Cutting Mills, another subclass, may differ in whether the knives are movable or fixed in the classifier configuration. The rotating element, including the rotating impeller, rotating screen, and oscillating bar, primarily distinguish Screening Mills. Tumbling Mills are distinguished by the grinding media used and whether the mill is vibrated, including ball media, rod media, and vibrating. While Separators are distinguished by their mechanism of operation, i.e., vibratory/shaker, centrifugal.

Table 5.7 Parameters

Process parameter		Formula by scale-up calculation	CASE—A			
			Range for exhibit batch	Scale-up calculation/trial/pre-PPQ scale	Actual recommended parameters for scale-up calculation	Remarks
Inlet air flow (cfm)	min	Min cfm required for large scale = min spray (small scale) × scale-up factor	300	300 * 1911 = 573	500–2500	Wider range given than the range derived from scale-up calculation. Range will be narrowed down based on process trials
	max	Max cfm for large scale = max spray (small scale) × scale-up factor	900	900 * 1911 = 1720		
Spray rate (g/min)	min	Min spray rate required for large scale = min spray (small scale) × scale-up factor	100	100 * 1911 = 191	100–1000	
	max	Max spray rate required for large scale = max spray (small scale) × scale-up factor	250	250 * 1911 = 478		

Table 5.8 Considerations

CPP	Impacted attributes	Approach (Higher scale)
Inlet/Product temperature (°C)	A pre-heated inlet/product bed ensures that the initial spray of coat does not over wet the fluidized mass	Dispersion to be sprayed when product temperature reaches a pre-established range. Temp. maintained through the entire process
Air flow (CFM)	Inadequate fluidization may result in lumps during spraying. Attrition may lead to generation of static charge and breakdown of the tiny pellets	Intentional stagewise titration of air flow: Low flow (and min fluidization) during preheating followed by planned ramp in air flow during spraying to support densification of mass over time
Atomization air pressure (bar)	Affects droplet size distribution and deposition onto the product ass. Coupled with spray rate, will influence the fluidization of mass and it results in lumps over time	Intentional and gradual increase of atomization air pressure titrated proportionally with gradual increase in spray rate
Spray rate (gm/min) and amount	Direct impact on total quantity of material deposited on the product mass. Dependent on fluidization and atomization of coating solution/dispersion. Impacts uniform deposition of coat	Low initial spray rate followed by stagewise ramp (ideally 2–4 ramps and can be adjusted based on weight buildup) of spray rate proportionate to increase in atomization air pressure and over time. Fixed, know excess coating dispersion/solution sprayed after characterization of process yields/efficiancies during development. No stop look go procedure. Move to next unit operation based on yields of preceding step
Gun position	Will have direct impact on particle size distribution	Gun position can be changed from bottom to middle & middle to top based on bowl load and intermittent increase in bed height during spray granulation

Screen Milling—e.g., Comil Selection and Studies

Comil [2] is effective for dry (deagglomeration—delumping) and wet milling of soft- to medium-hard materials. The comminution chamber consists of an impeller rotating at variable speed, imparting a compression or shear force inside a conical-shaped screen. The impeller imparts a vortex flow pattern to the feed material, and the centrifugal acceleration forces the particles to the screen surface and up the cone (360°) in a spiralling path.

The space between the impeller and the screen is adjustable. The size and shape of the screen holes, screen thickness, impeller configuration, and mill speed are important variables (Fig. 5.7).

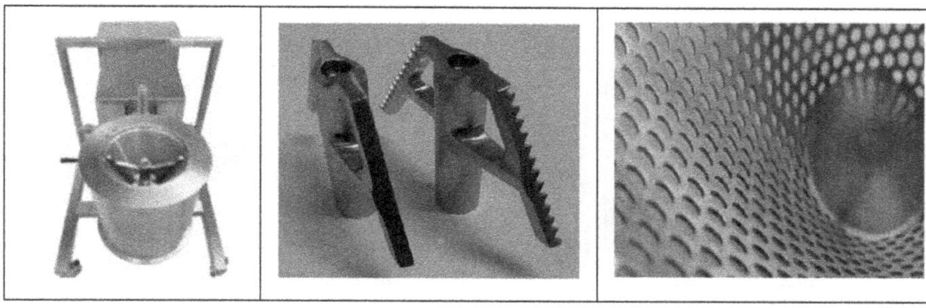

Fig. 5.7 Comil

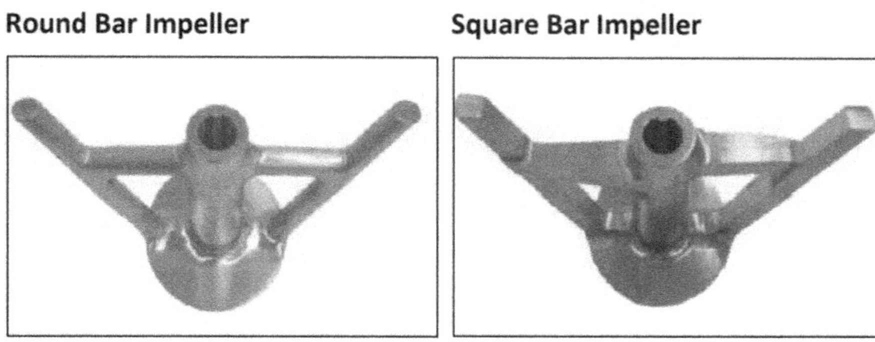

Fig. 5.8 Comil Impellers

The round bar impeller and square bar impeller are both impellers that can be used in a conical mill for dry and wet milling (Fig. 5.8). The main difference between the two is the "nip" angle, which refers to the angle between the impeller and the screen. The round bar impeller has a larger nip angle, allowing the particles to be "nipped" between the impeller and screen, resulting in quicker material processing. The square bar impeller, on the other hand, has a nip angle of almost zero, meaning the particles rely more on the centrifugal force imparted by the impeller to push them through the screen, resulting in slower processing of the material.

Co-Mill screens are designed with different hole shapes and sizes to suit specific applications. The R-Round Hole screen is used primarily for dry sizing or powder into liquid dispersion. The hole size ranges from 0.006 inches (0.15 mm) to 0.250 inches (6.35 mm) in diameter. The Q-Square Hole screen is used for wet (moist) product sizing and when larger particle sizes are required. The hole size ranges from 0.156 inches in height × 0.156 inches in length (3.96 mm × 3.96 mm) to 1.00 inches in height × 1.00 inches in length (25.4 mm × 25.4 mm). The S-Slotted Hole screen is used for breaking capsules before the "G" screen was introduced. The hole size ranges from 0.156 inches height × 0.380 inches (3.96 mm height × 9.65 mm length) to 1.00 inches height × 2.18 inches length (25.4 mm height × 55.37 mm length). The G-Grater Hole screen has a cutting edge on the holes. It is used for

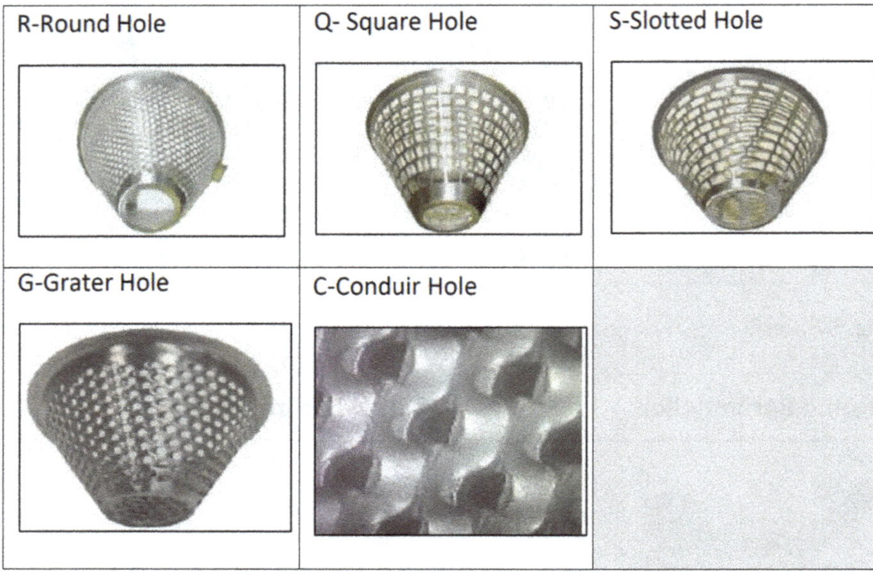

Fig. 5.9 Comil screens

sizing products that are harder or more ductile (with some elasticity). This screen is often used when a tighter particle size distribution curve is desired. The hole size ranges from 0.040 inches (1.00 mm) to 1.25 inches (31.75 mm) in diameter. The C-Conduir Hole screen also has a cutting edge on the holes. It is available in smaller holes than the "G" screen. This allows for finer milling with less chance of screen blinding. The hole size ranges from 0.006 inches (0.15 mm) to 0.039 inches (0.99 mm) in diameter (Fig. 5.9).

Scale-up variables are the parameters that need to be considered are (Table 5.9).

The selection of an impeller style depends on the specific requirements of the in-process quality attribute. A square-arm impeller with a greater "G" screen is effective for cutting the product, while a round-arm impeller is recommended for friable products that require a milled product with minimal fines. When selecting an impeller, it is essential to consider the product's characteristics, the desired outcome, and the operating conditions of the equipment. The shape, size, the number of blades, gap, and rotational speed of the impeller are all important factors that will affect its performance. The screen selection depends on the specific requirements of the in-process quality attribute. Greater screens, also known as "G" screens, are more effective at producing fine-cut granules, while round hole screens, also known as "R" screens, produce more fines. The "G" screens have a significant cutting edge that makes them better for fine cutting, while the "R" screens do not have this cutting edge which means they are less efficient for fine cutting. The selection of the screen hole size depends on the target particle size and the size reduction caused by the mechanical milling action of the Comil. To determine the required screen hole

Table 5.9 Scale-up variables

Variables	Scale dependent/ Independent
Screen size	Scale independent
Screen type	Scale independent
Impeller type	Scale independent
Impeller RPM	Scale-dependent

diameter size, the target particle size is multiplied by 1.5–2. For example, if the target particle size is 850 microns, the screen hole diameter should be 1700 microns (850 × 2). In this case, the closest available screen size would be a 062G with a hole diameter of 0.062 inches or 1575 microns. It is possible to estimate the particle size range resulting from a specific screen based on the hole diameter size. By dividing the hole diameter by 2, we can estimate the minimum particle size, and by dividing it by 1.5, we can estimate the maximum particle size. For example, if we have a 079G screen with a hole diameter of 0.079 inches (1016 microns), we can estimate that most of the product will be in size range of 1016–1346 microns (0.040–0.053 inches). However, it is important to remember that different materials may mill differently, but this rule is typically true.

A scale-up example is of mill speed from U10/197 to U20/194.

A. Scale—U10/197, Impeller tip dia—110 mm, Std rev per min—2400, Operating speed—1500
B. Scale—U20/194, Impeller tip dia—194 mm, Std rev per min—1400, **TBD**

The mill speed can be scaled up using one of the three options:

1. By keeping the same tip speed across all scales. The tip speed is the speed at which the tip of the impeller travels and is measured in feet per minute.

   Tip speed at lower scale $(V_1) = 2 * \pi * R_1 * N_1$
   Units for tip speed are mm per minute.
   $R_1 =$ lower scale impeller radius in mm
   $N_1 =$ lower scale impeller rotation speed in rpm
   Tip speed at higher scale $(V_2) = 2 * \pi * R_2 * N_2$
   $R_2 =$ higher scale impeller radius in mm
   $N_2 =$ higher scale impeller rotation speed in rpm
   Therefore, $(V_1) = 2 * \mathit{3.14} * \mathit{(110/2)} * \mathit{1500} =$ 518100 mm/min
   Tip speed should be matched in higher scale for scale-up.
   $(V_2) = 2 * \pi * R_2 * N_2\ c$
   $N_2 = 518100/2 * \pi * R_2 =$ **851 RPM**

2. By matching the Froude Number across all scales. A Froude number is a dimensionless number that is used to describe the ratio of inertial forces to gravitational forces and is used to predict flow patterns in a fluid system.

Froude number at lower scale is $\mathrm{Fr}_1 = N_1^2 * D_1/g * 3600$
Froude number at higher scale is $\mathrm{Fr}_2 = N_2^2 * D_2/g * 3600$
N = Impeller rotations (RPM)
D = Impeller diameter (meters)
$g = 9.8$ m/s^2
$\mathrm{Fr}_1 = 1500 * 1500 * (110/1000)/9.8 * 3600 = 7.02$
Froude number should be matched across the scales
$\mathrm{Fr}_2 = N_2^2 * D_2/g * 3600 = 7.02$
$N_2 = \sqrt{7.02 * g * 3600/D_2}$ = **1130 RPM**

3. By matching the standard impeller rotations (SIR) across all scales. The standard impeller rotation is the number of rotations per minute of the impeller.

Higher scale RPM based on SIR formula is $N_2 = N_1 * S_2/S_1$
N_1 = lower scale rpm
N_2 = higher scale rpm
S_1 = lower scale SIR
S_2 = higher scale SIR
$N_2 = 1500 * 1400/2400$ = **875 RPM**

The choice of criteria will depend on the specific application and the desired outcome.

Blending and Mixing Technologies

Blending is the process of mixing various bulk powder ingredients to achieve homogeneity and uniform distribution of the mixture [3]. The quality of the resulting blend depends on several factors, such as the types of blenders or mixers used, the flowability of the powder during the blend cycle, and the mechanics of blending. The mechanics of blending mainly involve the principles of diffusion, convection, and shear. Diffusion is the process by which particles of different sizes and densities move and intermingle due to random thermal motion. Convection is the movement of particles caused by fluid flow, in this case, air fluidization. Shear is the mechanical force applied to the particles to break down any agglomerates or clusters that may have formed during the blending process. There are various types of blenders or mixers used in the pharmaceutical industry, such as V-blenders, double-cone blenders, and tumble blenders (Fig. 5.10). The blender choice will depend on the desired final blend uniformity, blend cycle time, and the type of blended ingredients (Table 5.10). It's important to note that sampling (Fig 5.12), blend homogeneity and blend uniformity are critical parameters for any blending process and can be achieved by the proper selection of blender type, blend cycle time, and flowability of powder during blending.

Diffusion, also known as dispersion, is the process by which particles within a powder blend redistribute and change position relative to one another due to random

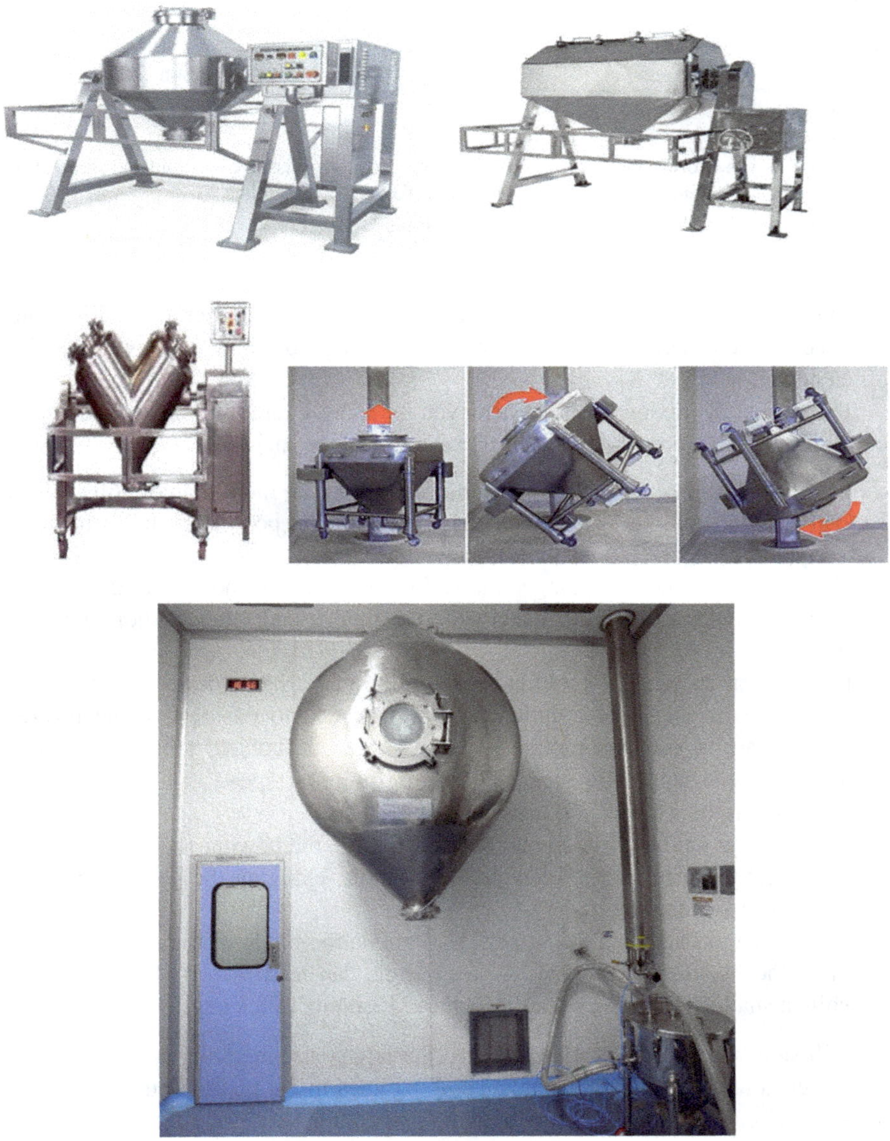

Fig. 5.10 Blenders

thermal motion. This process helps to achieve homogeneity and uniformity in the blend by ensuring that each particle has an equal chance of coming into contact with every other particle. Double cone blenders, Octagonal blenders, V-shell blenders, and bin blenders are examples of diffusion blenders, also known as tumbling or rotating shell blenders. These blenders work by rotating the powder blend in a container, causing the particles to move randomly and come into contact with one

Table 5.10 Guideline for mixing

Direct blending	Preblending (Nongeometric)	Preblending (Geometric)	Solvent addition
>5–10% active	1–5% active	<1% active	<0.1% active
Straight mixing	Use (drug fraction)^{½}-Preblend	1:1 (Drug: Exc) 1:1 (Mix: Exc)...	Dissolve in liquid and spray/coat/granulate
Adequate	Good Method	Cumbersome	For low dose

another through diffusion. The rotation also causes the particles to be lifted and dropped, which helps to prevent particle segregation and improve blending efficiency.

The degree of mixing/blending achieved by using a tumbling mixer in carrying out a mixing operation is dependent on several factors, including:

- The fill-up volume: The fill-up volume should not be more than 50–60% of the total blender volume to allow enough space for the powder particles to move freely and come into contact with one another.
- The residence time: The longer the residence time, the more time the powder particles have to come into contact with one another and achieve a homogeneous blend.
- The rotation speed: The rotation speed of the mixer has a direct impact on the blending efficiency. An optimum speed is needed to ensure that the powder particles are in constant motion and can come into contact with one another. However, increasing the speed above the optimum speed can cause powder adhesion on the walls of the mixer.
- The charging method: The method used to charge the powder into the mixer can also affect the blending efficiency. For example, charging the powder in layers can lead to segregation.
- The inclination angle of the mixer: The inclination angle of the mixer can also affect the blending efficiency. A steep angle can lead to particle segregation, while a shallow angle may lead to reduced blending efficiency.

Diffusion mixers are best suited for free-flowing, nonsegregating powders and are widely used for blending dry powders and granules in the pharmaceutical industry because of close quality control on batch operation and an effective diffusive mechanism. They can also be used to accomplish wet blending as in the case of a modified V-blender (V-blender with an intensifier bar). They find use in gentle mixing of friable solid material and can also be used for mixing powders of different densities. However, diffusion mixers have some limitations too. Large differences in particle densities or wide particle size distributions of materials can lead to segregation and highly cohesive materials are difficult to mix.

Convection mixers are a type of mixing equipment that uses impellers (blades or agitators) to circulate and blend materials. The shape of the vessel and the geometry of the impeller primarily distinguish the different subclasses of convection mixers.

Fig. 5.11 Ribbon blender

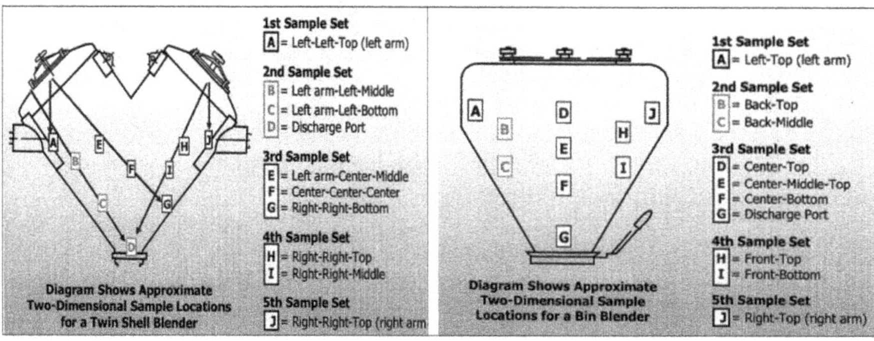

Fig. 5.12 Sampling plan examples

Ribbon blenders (Fig. 5.11) use a horizontal U-shaped ribbon agitator to mix materials. Orbiting screw blenders use a screw-shaped impeller that rotates on an offset shaft to mix materials.

Planetary blenders use a beater-shaped impeller that rotates on an offset shaft and orbits around a stationary center post to mix materials. Forberg blenders use a rotating drum with internal paddles to mix materials. Horizontal double-arm blenders use two horizontal agitators rotating in opposite directions to mix materials. Horizontal high-intensity mixers use high-speed, high-shear impellers to mix materials. Vertical high-intensity mixers use high-speed, high-shear impellers to mix materials in a vertical vessel. Diffusion mixers (tumble mixers) use an intensifier/agitator to mix materials by tumbling them in a rotating drum. Pneumatic mixers are a type of mixing equipment that uses compressed air to mix materials. They may differ from one another in terms of vessel geometry, air nozzle type, and air nozzle configuration. Pneumatic mixers are typically used for low-viscosity liquids, powders, or granules. Pneumatic mixers are generally considered less expensive and more energy efficient than mechanical mixers. Still, they may not be suitable for high-viscosity materials or materials sensitive to air entrapment.

Table 5.11 Scale-up/down blender speed and blending time

Scaleup parameters	Scale-up procedure
Batch size	Batch size will be calculated on higher scale by keeping % occupancy constant
Blender speed and blending time	*Option 1: Tip speed matching* Tip speed m/min = $2\pi * r * R$ $r = h/2$ where h is the height of the blender R = RPM of the blender Match tip speed and keep the # of revolutions constant by varying time *Option 2: Study the number of revolutions range* Select the blender RPM based on the blend properties Perform the blend uniformity at the different total number of revolutions (TNR) e.g., RPM = 10 TNR range selected: 100, 150 & 200 (10 × 10, 10 × 15, and 10 × 20) for pre-lubrication TNR range selected: 30, 50 & 70 (10 × 3, 10 × 5, and 10 × 7) for lubrication

Fixing the Batch Size

Fixing the batch size is a preliminary step in scale-up, and it is calculated on a higher scale by keeping the percentage occupancy constant, taking into consideration the working volume, and using an equation to calculate the batch size (Table 5.11):

$$B = V \times D$$

[where B = Batch size in kg, V = working vol. of the container, D = product bulk density (g/cc)].

Compression Technologies

Tablet press subclasses can be distinguished by the method used to deliver powder blends to the die cavity (Fig. 5.13). These methods include:

- Gravity, where powders are delivered without mechanical assistance
- Power-assisted, where powders are delivered with mechanical assistance
- Centrifugal, where powders are delivered by rotational forces
- Compression coating, where a tablet core is formed in one location and then coated with an outer layer of material in another location
- Paddle, where powders are delivered by paddles.

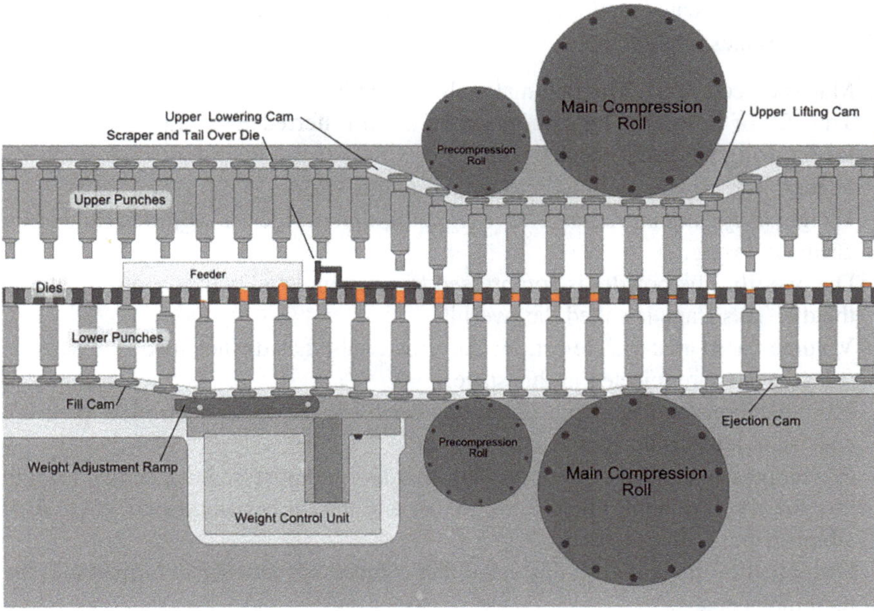

Fig. 5.13 Tablet compression

Gravity feed tablet presses use the force of gravity to feed the powder into the die cavity [4]. They are typically used for small-scale production and have a lower output capacity compared to other types of tablet presses. Power-assisted tablet presses use mechanical assistance, such as a screw feeder or auger, to feed the powder into the die cavity. They are typically used for medium to large-scale production and have a higher output capacity than gravity feed presses. Centrifugal tablet presses use rotational forces to feed the powder into the die cavity. The powders are fed into a spinning drum, which forces the powders into the die cavity through centrifugal force. They are typically used for high-volume production and can have very high output capacities. Compression coating tablet presses are used to apply an outer layer of coating material to a preformed tablet core. This method is often used to add a protective coating or to change the appearance of the tablet. Paddle tablet presses use paddles to feed the powder into the die cavity. They are typically used for small-scale production and have a lower output capacity compared to other types of tablet presses. They are mostly used in R&D and lab scale applications.

The tablet compression process involves compacting powders into a solid form using a machine called a tablet press. The physics of powder compaction were developed in the 1950s, which led to the design of more efficient presses, improved feed systems, multistage compression, advanced automation and control, high-speed reject mechanisms, and the ability to create multitechnology tablets such as bilayer, multilayer, core coated, and plug-in-tablets. A rotary compression machine consists of several stages, including:

- Material Feed Zone: The raw materials are fed into the machine in this stage.
- Transfer of Powder: The raw materials are transferred from the hopper to the feeder using a transfer mechanism.
- Feeder: The powder is filled into the die in this stage.
- Weight Adjustment Zone: The weight of the powder is adjusted to ensure accurate dosing.
- Die overfill: The powder is compressed to a certain level before being filled into the die. This stage is called die overfill.
- Volume-based material rejection: Any materials that do not meet the volume requirements are rejected in this stage.
- Fill depth control: The depth of the powder fill in the die is controlled to ensure consistent tablet thickness.
- Precompression and compression: The powder is compressed to a certain level in this stage before being filled into the die, and then it is compressed to the final tablet shape.
- Formation of the tablet: In this stage, the compressed powder is formed into the final tablet shape.
- Ejection and scrape off: The tablet is ejected from the die, and any excess powder is scraped off.

The tablet press can be operated in a manual or automatic mode. Modern tablet presses are also equipped with computer control systems that allow for precise control of the compression process and the ability to store and retrieve production data (Table 5.12).

During tablet compression, powders undergo changes in physical properties, including consolidation, compaction, and deformation. Consolidation is the reduction of bulk volume and displacement of air. It occurs during the initial stages of compression. Compaction and deformation refer to the further reduction of volume and the physical changes that occur in the powders. Three types of deformation can occur during compression: elastic, plastic, and brittle fracture. Elastic deformation is reversible and occurs when a material is compressed and returns to its original volume upon removal of the load. All materials undergo elastic deformation to some extent, but the extent depends on the nature of the material. During the initial

Table 5.12 Machine controls

Operating parameters	Machine parameters
Turret speed	IDF paddle configuration
IDF speed	Fill cam size
Fill depth	Roller tonnage
Insertion depth	Peak punch tip force
Edge thickness	Ejection cam size
Precompression force	Tapered die
Compression force	Keyed uppers/lowers

compression stages, microcrystalline cellulose (MCC) undergoes elastic deformation. Plastic deformation is nonreversible and results in a vicious "flow" of the material. It is a time-dependent process, and the degree of deformation depends on the dwell time. Fibrous excipients, starch, lubricants, waxes, and hydrophobic materials typically undergo plastic deformation. High quantities of these materials show a plateau in hardness with increased compression force. A brittle fracture occurs when the elastic yield strength of a material is exceeded, and the shear strength is much greater than the breaking strength. Larger particles are sheared and broken into smaller particles, typically in the later stages of compression. Lactose, dicalcium phosphate, and mannitol are examples of materials that undergo brittle fracture. Adequate lubrication is required to account for the increase in in-die surface area and die wall ejection.

Decompression is the process of removing compression force from a compressed material. The rate of removal of compression force can have an impact on the recovery of the material. Immediately after compression, the material may experience elastic recovery in the axial direction within the die. If the elastic and recovery forces are high, capping, or lamination may occur. This can result in the actual tablet thickness being greater than the in-die tablet thickness. Ejection is the process of removing a tablet from the die, during which the tablet remains in the die until it is fully ejected. It typically occurs in three stages: peak force to break the tablet's adhesion to the die wall, the force required to push the tablet up the die wall, and declining forces as the tablet are fully ejected out of the die. Capping during ejection can occur due to high elastic forces and recovery in the axial direction, and sudden recovery in the radial direction upon ejection. Scrape off is the process that occurs immediately after ejection in which the tablets are released from the lower punches. The typical forces used during scrape off are 2N or less. Issues may occur with traditional design, so modification can be done to avoid the problems. Some of the common issues are lower tablet surface may pick if punch adjustment is not proper, and shape and orientation may cause breakage at the scrape-off location.

Flow control from the hopper into the feed frame involves regulating material flow with a calibrated gate to prevent flooding and retro-flow of air. A level sensor can also be installed to monitor the material level in the hopper. A vibrator may also be used to aid in the flow of material, but it can also hinder the process if not properly adjusted. A gravity feeder is an open design that relies on gravity to fill the die. The speed of the press may be limited by the rate at which the die is filled. This type of feeder is suitable for free-flowing powders with high bulk density and some segregation-prone blends. It is also suitable for lubrication-sensitive, low specific surface area, high magnesium stearate blends, and low tablet-weight products. However, it may not be suitable or appropriate for highly potent products because the design of the gravity feeder is open, which may allow cross-contamination. The force feeder uses a dual or triple-chamber design with a rotating blade that pushes material into the die. The first chamber always rotates in the opposite direction to the turret. The shape of the blade may vary between vendors. This feeder is suitable for high-speed tablet operations and large tablet weights with high turret speeds. However, it is important to carefully monitor the residence time of the blend to match the blend usage rate, as it may cause over-lubrication over time. Flooding, leaking, and recirculation may also become excessive. The feeder speed should be kept as low as possible to minimize these issues. The weight regulation zone in the tablet press is where the weight of the tablets is regulated. Volumetric powder fill can be affected by variations in size distribution, density, and feed rate into the die. These variations can cause fill depth variations, which can affect the weight of the tablets. The correct cam size is critical for obtaining optimal over-fill. A small cam may not provide enough over-fill, causing weight variation, but if the flow is good, high speeds can be achieved. A large cam may provide too much over-fill, causing weight variation at high speeds due to inadequate weight regulation. Over-fill may also cause a reduction in powder strength for friable granules, over-lubrication, and segregation for fragile blends.

Scale-up correlations are used to predict how a process will perform at a larger scale based on data from a smaller scale. In compression machine speed, one of the key factors that determine the speed is the dwell time (DT). The dwell time is the time the material spends in the compression chamber before it is compressed. It is critical to ensure that the dwell time remains constant during the scale-up process, as variations in dwell time can affect the tablet's weight and quality. The compression machine speed can be adjusted accordingly to achieve the optimal dwell time for a specific process.

$$\mathrm{DT} = \mathrm{PHF} \times 60{,}000/\pi \times \mathrm{PCD} \times N$$

where
PHF (Punch head flat) = 12.7 mm of B tooling; 18.23 mm for D tooling
N = # of rotations per min of turret
PCD = Pitch circle diameter of turret (mm)
DT (msec) = Dwell time in milli seconds

In a case study of tablet compression for drug A, the problem condition is a lower assay value.

Process 1: A blend was compressed in 26 station compression machines with 4 sets of punches at 20, 25, 30, 40, and 50 RPM; Assay results are as follows:

RPM	20	25	30	40	50
Assay (%)	79.8	80.6	81.5	83.2	85.7

Process 2: Same process with 13 set of punches at 20, 25, 30, 40, and 50 RPM; Assay results are as follows:

RPM	20	25	30	40	50
Assay (%)	99.5	99.8	100.1	100.4	100.2

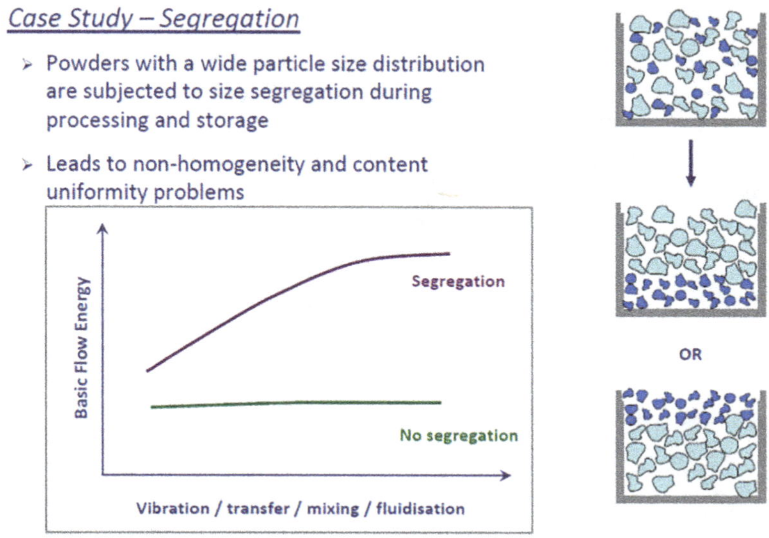

Fig. 5.14 Case study

FT4 Rheometer data:

SI (Stability index): 1.28 (Value very much greater than 1 indicates a powder with high segregation tendency provided the particle size distribution is wide)

PSD Data for Batch 1:

60# retain = 33%, 100# pass = 46% (This indicates a wide particle size distribution)
SE (Specific Energy): 5.52 (SE value of greater than 5 indicates a cohesive powder)

So, the powder is moderately prone to sticking inside the feeder.
Assay result of 100# separated blend:

Assay of 100# passed blend = 75%
Assay of 100# retained blend = 105%

The inference from the information provided is that the lower assay value in the tablets produced is likely due to the presence of finer particles in the blend that are separated inside the feeder and filling the dies (Fig. 5.14). These finer particles may have a lower assay than the rest of the blend, leading to a lower overall assay in the tablets. To mitigate this issue, the blend retention time inside the feeder should be reduced to match the retention time used with a full set of punches, where the assay is almost 100%. By reducing the retention time, the finer particles will have less time to separate and fill the dies, resulting in a higher overall assay in the tablets. It is important to note that this is a hypothesis. It is important to conduct experiments and monitor the results to determine the exact reason for assay loss and the best way to mitigate it.

Calculation of Blend retention time:

Let the weight of the tablet be W mg and the Turret speed be N RPM
For 4 sets of punches, weight of blend consumed per unit time, $(W1) = 4 * N * W$ mg/min
For 13 sets of punches, weight of blend consumed per unit time, $(W2) = 13 * N * W$ mg/min
Retention time inside feeder (in minutes)

$$(T1 \text{ or } T2) = 1000M/(W1 \text{ or } W2)$$

M = Maximum capacity of feeder in gm = 400 g (approx.) for CTX 26 station

For 1 mg, and 2 mg unit weight $(W) = 60$ mg

Turret RPM	*W*1 (g/min) 4 sets	*W*2 (g/min) 13 sets	*W*2 (g/min) 32 sets	*T*1 (min): 4 sets	*T*2 (min): 13 sets	*T*2 (min): 32 sets
20	4.8	15.6	38.4	**83.3**	**25.6**	**10.4**
25	6.0	19.5	48	**66.7**	**20.5**	**8.3**
30	7.2	23.4	57.6	**55.6**	**17.1**	**6.9**
40	9.6	31.2	76.8	**41.7**	**12.8**	**5.2**
50	12	39	96	**33.3**	**10.3**	**4.2**

Feeder capacity = 400 g (approx.)
$(W1) = 4 * N * W$ mg/min
$(W2) = 13 * N * W$ mg/min
$(T1) = 1000M/W1$
$(T2) = 1000M/W2$

The case study revealed that lower assay values in Drug A tablets were caused by the segregation tendency of the blend, which resulted in the separation of finer particles. To mitigate this issue, it was determined that it is necessary to minimize the residence time of the blend within the feeder. This can be achieved by using a full set of punches and increasing the turret speed while adjusting the force feeder RPM to achieve optimal results.

Encapsulation Technologies

Encapsulator subclasses can be distinguished by the method used to introduce materials into the capsule. These methods include Auger, Vacuum, Vibratory, Dosing disk, Dosator types. Auger encapsulators use a rotating auger to feed material into the capsule. This method is suitable for powders or granules. Vacuum encapsulators use a vacuum to suck material into the capsule. This method is suitable for liquids or pastes. Vibratory encapsulators use a perforated plate that is vibrated to force material into the capsule. This method is suitable for powders or granules.

Dosing disk encapsulators use a tamping method to fill the capsule. A bored disk is used to compress the material into the capsule. This method is suitable for powders or granules. Dosator encapsulators use cylindrical tubes fitted with pistons to feed the material into the capsule. This method is suitable for liquids or pastes. These methods have their own advantages and disadvantages and are chosen based on the type of material that needs encapsulation and the desired output.

Challenges

Typical Issues faced during technology transfer:

- Material factors: Changes in API quality: impact of cost-containment initiatives on quality: changes in critical excipient quality, lack of appropriate performance-related specifications: changes in packaging components/parameters, blister seal integrity, desiccation capacity.
- Equipment factors: Type and design of equipment, hopper and feed frame designs, pitch angles of blades, charging, discharging and shipping mechanisms, testing of retrofits for process control measurements, lack of automation: manual errors, equipment capability to measure some critical parameters is different across scale, lack of process analytical tools to determine end-point and control process parameters on-line to achieve particular quality attributes, lack of established correlation to scale-up critical process parameters.
- Human factor: Multiple operators in different shifts, training and education of workforce, underlying science and rationale, clarity of instructions.
- Documentation: Technology transfer documents, ownership: transfer, maintenance, live document, level of details, relation to contents of filing, clarity/comprehension, science and rationale.
- Systems factors: Powder/granulation transfer systems, flexibility needed (for learning) during prevalidation and validation lots, utilization of surrogates to ensure appropriate performance.
- Environmental factors: Design of the facilities, transfer mechanisms, HVAC: Temp/humidity controls, conditioning of inlet air, in-process storage—length and conditions.

References

1. FDA (1995) SUPAC-IR: immediate-release solid oral dosage forms: scale-up and post-approval changes: chemistry, manufacturing and controls, in vitro dissolution testing, and in vivo bio-equivalence documentation. US FDA.
2. Kotamarthy L, Metta N, Ramachandran R (2022) Understanding the effect of granulation and milling process parameters on the quality attributes of milled granules. Processes 8:683
3. Alyami H, Dahmash E, Bowen J, Mohammed AR (2017) An investigation into the effects of excipient particle size, blending techniques and processing parameters on the homogeneity and content uniformity of a blend containing low-dose model drug. PLoS One 12:20170613
4. Lachman (1991) Theory & practice of industrial pharmacy. Wolters Kluwer

Chapter 6
Qualification, Continued Process Verification, and Lifecycle Management

Abstract Process validation lifecycle qualification and continued process verification (CPV) stages are essential in the manufacturing of pharmaceutical products. Stage 2 involves the qualification of equipment, utilities, facility, and system designs, as well as the assessment of the manufacturing process to ensure consistent quality products during commercial manufacturing. This is crucial before the commercial distribution of a drug product. In Stage 3a, a substantial body of data is used for statistical evaluation to gain a deeper understanding of the product after the new product launch. Stage 3b involves routine CPV and trending of critical process parameters and critical quality attributes within established alert limits. The effective application of the post approval change management protocol (PACMP) defined in ICH Q12 is a great organizational strategy for future product lifecycle change management, providing a tool to make predetermined changes without impacting the product supply.

Keywords Qualification · Process Validation · Lifecycle Management · PACMP · PPQ · Stage 2 · Stage 3a · Stage 3b · Continued Process Verification

Introduction

The goal of Stage 2 is to ensure that the process design developed in Stage 1 can produce consistent quality products during commercial manufacturing. This includes the qualification of equipment, utilities, facility, and system designs (Stage 2a) [1] and the assessment of the manufacturing process (Stage 2b). Completing Stage 2 is crucial before the commercial distribution of a drug product. In Stage 2b, the products are primarily tested in a way that simulates commercial manufacturing settings while collecting additional data to statistically justify the use of the manufacturing process for continued commercial manufacturing. Stage 2b is defined as confirming that the manufacturing process as designed is capable of reproducible commercial manufacturing, as per US FDA guidance on Process Validation.

A. B. Pazhayattil et al., *Technology Transfer*, AAPS Introductions in the Pharmaceutical Sciences 10, https://doi.org/10.1007/978-3-031-32192-4_6

Stage 2a

Before commercial production for distribution, it is necessary to qualify all facilities, utilities, and equipment used for their design, installation, and operation for intended use, cleaning, maintenance, and performance. This includes facility design and commissioning. Qualification activities for facilities, utilities, systems, manufacturing, and packaging equipment include risk management to determine the impact on product quality attributes. The ASTM 2500 [2] may be used to assess the criticality of each piece of equipment, utility, and system on the potential to impact product quality attributes. To ensure proper operation and maintenance of the system, necessary documents and procedures must be in place before production. Qualification is carried out by trained personnel and in accordance with predetermined and approved qualification protocols and standard procedures. The results of qualification activities should be recorded in approved qualification reports. The four stages of qualification that are needed to determine if processes are suitable are

- Design Qualification (DQ)
- Installation Qualification (IQ)
- Operational Qualification (OQ), and
- Performance Qualification (PQ)

All utilities and equipment should be maintained in a qualified state and periodically reviewed. Any changes made should be assessed through appropriate change control systems, and the impact of the change should be assessed per quality risk management principles to determine the tests required to maintain the qualification status. To ensure consistency from batch to batch, it is important to validate. If not, all qualification stages are required for new systems, and it is necessary to provide documented impact assessment and scientific justification to support the decision. Systems such as computerized systems, water purification systems, steam systems, production, and quality control equipment and instruments should be qualified. When third parties conduct some stages of qualification, it is the manufacturer's responsibility to ensure that the qualification is completed in accordance with good manufacturing practices (GMPs).

Process Performance Qualification Prerequisites

There are several prerequisites that should be met before starting Stage 2b PPQ:

- The design of the facility, equipment, and utilities should be qualified (Stage 2a)
- The process design should be finalized, and all process parameters identified and controlled
- All procedures, such as cleaning, maintenance, and testing, are developed and validated

- Validation master plan (VMP) should be in place outlining policy, organization, planning, scope, and stages that impact the product quality
- The appropriate personnel should be trained and in place to execute the PPQ

It is important to note that the prerequisites [3] may vary depending on the product, process, and regulatory requirements.

PPQ Protocol and Reports

When creating a protocol for Stage 2b PPQ (process performance qualification), it is important to develop a sound plan for sampling and testing. This plan should consider factors such as dosage strengths, drug mechanism of action, release mechanism, absorption, and manufacturing process steps. Parameter set points and ranges that were not established during Stage 1 should be established during the Stage 2b study, including dosing batch size, compression speed, coating pan load range, and coating operating guidelines. A minimum number of consecutive blend batches and compression and coating validation batches should be manufactured according to Master Manufacturing Records to meet the requirements of Stage 2b. The testing plan should evaluate whether the process consistently produces a product that meets predetermined specifications for chemical and physical properties. The protocol should also include a reference to the approved Master Manufacturing Records, materials and specifications, equipment to be used and its qualification and calibration status, requirements for training, facility, utility, and environmental conditions, and completion of a PPQ prerequisite checklist.

Determining the Number of Batches

The 2011 FDA guidance states that Stage 2b PPQ activities must be based on sound scientific reasoning, a good understanding of the product and process, and an appropriate demonstration of process control. The number of samples taken should be sufficient to give statistical confidence in the quality of the product both within and between batches. Therefore, the number of Stage 2b batches that need to be tested should be determined by analyzing the within- and between-batch variability [4].

The optimal strategy is to determine the minimum number of batches needed for a projected confidence interval of the product's critical quality attributes to fall entirely within the desired specifications. This means that, based on available information, the number of batches that will provide enough data for a statistically confident conclusion about the product's critical quality attributes should be identified. To meet current specifications, the tested mean of a product quality attribute should be as close as possible to the center of the specification, and its standard deviation

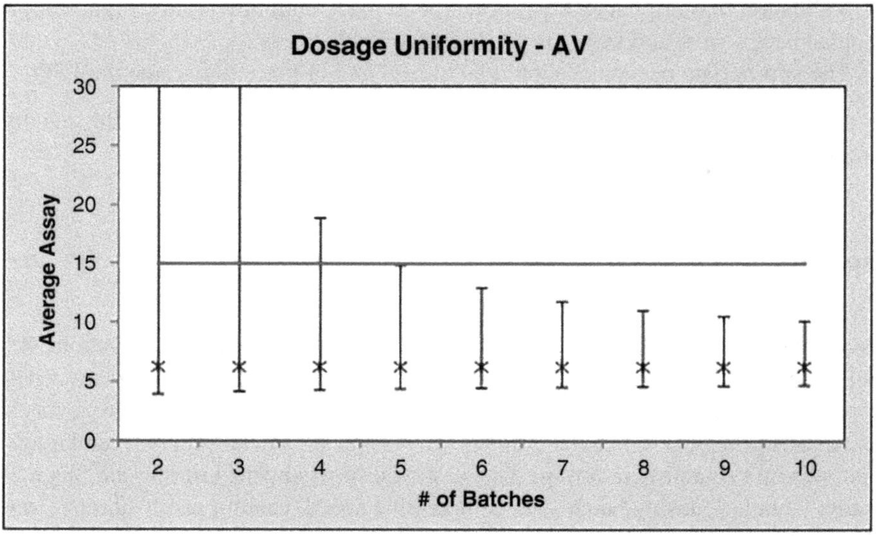

Fig. 6.1 Dosage uniformity

should be as small as possible, assuming a normal distribution. A confidence interval is then estimated for each number of potential PPQ batches based on previously collected product-specific data (i.e., the magnitude of within-batch statistics) and historical evidence of batch-to-batch variability for similar products (i.e., between batch). The projected number of PPQ batches is determined where the entire confidence interval falls within the specification limits. This approach is illustrated below for the dosage uniformity AV quality attribute (Fig. 6.1).

The form of the equation is dependent on the specific quality attribute; comparable derivations can be accomplished for other attributes, such as assay and dissolution. It should be noted that the number of PPQ batches estimation does not supplant the need to produce the PPQ batches or review the data generated from these batches. The total or overall variability of a process can be represented as a summation of individual component variation. This may be mathematically denoted as

$$S^2{}_{\text{total}} = S^2{}_{\text{batch}-\text{batch}} + S^2{}_{\text{intra}-\text{batch}} + S^2{}_{\text{sampling}} + S^2{}_{\text{analytical}} + \ldots.$$

The total variation is composed of variation derived from batch-to-batch, intra-batch, sampling, and analytical variability sources. Generally, the Stage 1 process design assesses most variation sources with the notable exception of the batch-to-batch (between or inter-batch) variability. Thus, data from Stage 1 provides a reasonable measure of product intra-batch performance. However, it is impossible to assess the batch-to-batch variability until several batches of products are produced and analyzed. To approximate this component, it is reasonable to assert that a similar process/product will exhibit similar batch-to-batch characteristics (Fig. 6.2). As

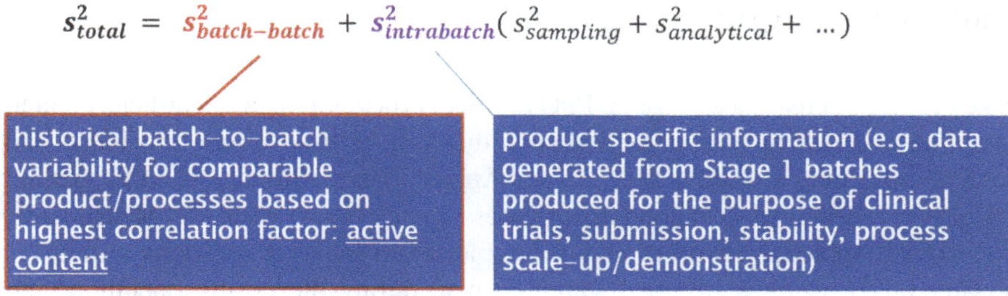

Fig. 6.2 Variability

such, tabulated evidence from historical records can provide a good estimate. Therefore, the number of Stage 2b PPQ batches required is the number of batches when the projected "best estimate" confidence interval of the product quality attribute measurements (which is a combination of the CI of the process mean and the CI of the process standard deviation) resides completely within the specification range.

An essential element in determining the number of PPQ batches needed is the batch-to-batch variability. Before the PPQ campaign, this factor is not established. However, data from similar campaigns provides an indication of the expected batch-to-batch variability. The magnitude of batch-to-batch variability may depend on various factors, such as the API content or product label claim. To gain insight into batch-to-batch variability, historical dosage uniformity and dissolution data from previous batches and molecules must be compiled. The batch-to-batch variability is then extracted from each campaign by separating intra-batch variability from overall campaign variability. The distribution of each campaign's batch-to-batch standard deviation results in a typical distribution profile expected for a collection of standard deviation data. Summary data from the historical evidence is stored in reference tables and used as an approximation of the batch-to-batch component of variation used to justify the number of batches to be evaluated during PPQ, providing reasonable confidence that the evaluated process is robust.

Once PPQ batches are manufactured and tested, it is wise to compare the estimation of batch-to-batch variation with the truly observed PPQ batch-to-batch variation. It should be noted that both product-specific information and historical batch-to-batch process information may vary significantly among different manufacturing facilities (due to personnel, operation, process, equipment, raw material, and other factors). The sources of variation for other types of manufacturing technologies will differ. To statistically justify the number of validation batches to be produced, the company should understand the variation they observe from various processes based on their historical data. Several methodologies are available that address the challenge of justifying a statistical model for determining a sufficient number of batches.

Sampling Strategies

A key aspect of creating a Stage 2b PPQ protocol is developing a scientifically sound sampling and testing plan. One approach that can be used to confirm blend and content uniformity at scale is the Grouped Area Variance Estimate (GAVE) method [5], which is designed to fit solid dose processes and comply with ASTM standards. The GAVE approach involves taking samples from multiple locations within the powder blend, using a stratified sampling plan during the dosing operation, and testing a minimum of seven units from each of the 20 sample locations. This results in a total of 140 units for dosage uniformity testing and helps to ensure that the process consistently produces a product that meets predetermined specifications for chemical and physical properties (Fig. 6.3).

Although traditionally 10 locations are deemed adequate to map tumbling-type mixers, 12 locations are recommended based on the geometry of bin blenders. There are, therefore, sufficient samples to determine within- and between-location variability. This is accomplished by a grouping of locations within an area. Samples for dosage uniformity testing are taken throughout the semi-continuous dosing process. A stratified sampling plan is followed during a dosing operation. The first sample is taken at the start, and the remaining samples are taken at equal intervals until the end of the process, with sampling points at no more than 5% of the batch. This results in a total of 20 strata samples of dosages representing the entire dosing run.

Stratified sampling increases the confidence of the uniformity of the batch as sample locations target problematic areas prone to potential segregation. A risk-based sampling plan that includes additional sampling points at the beginning and the end of the run may be applied based on product/process knowledge. Random variation is expected throughout a population; therefore, CU results are expected to be a normal distribution with no special cause variation. A minimum of seven units from each of the 20 locations are required to be sampled for testing. For statistical rigour, 20 sample locations are sufficient. Three random samples are tested from each of the 20 strata resulting in 60 units from each blend for Tier 1 testing and a total of 140 (20×7) units for Tier 2 dosage uniformity testing. Variance component analysis (VCA) allows for quantifying sources of variability across different levels of units from a sampling scheme. VCA analysis is conducted to determine the potential source of variability: either within area and/or between areas. High between-area variance could indicate poor mixing, resulting in nonuniformity within the blender and the potential for segregation. High within-area variance could indicate sampling bias or analytical errors. In the sampling plan proposed, single testing per location is performed with multiple tests per area, hence VCA is possible. Adequate mixing is directly assessed since it is more likely that mixing issues will appear as a difference between the top, middle, and bottom areas of the blender than between locations lying potentially in the same area. When between-areas variability is observed, it indicates uniformity issues related to process/product.

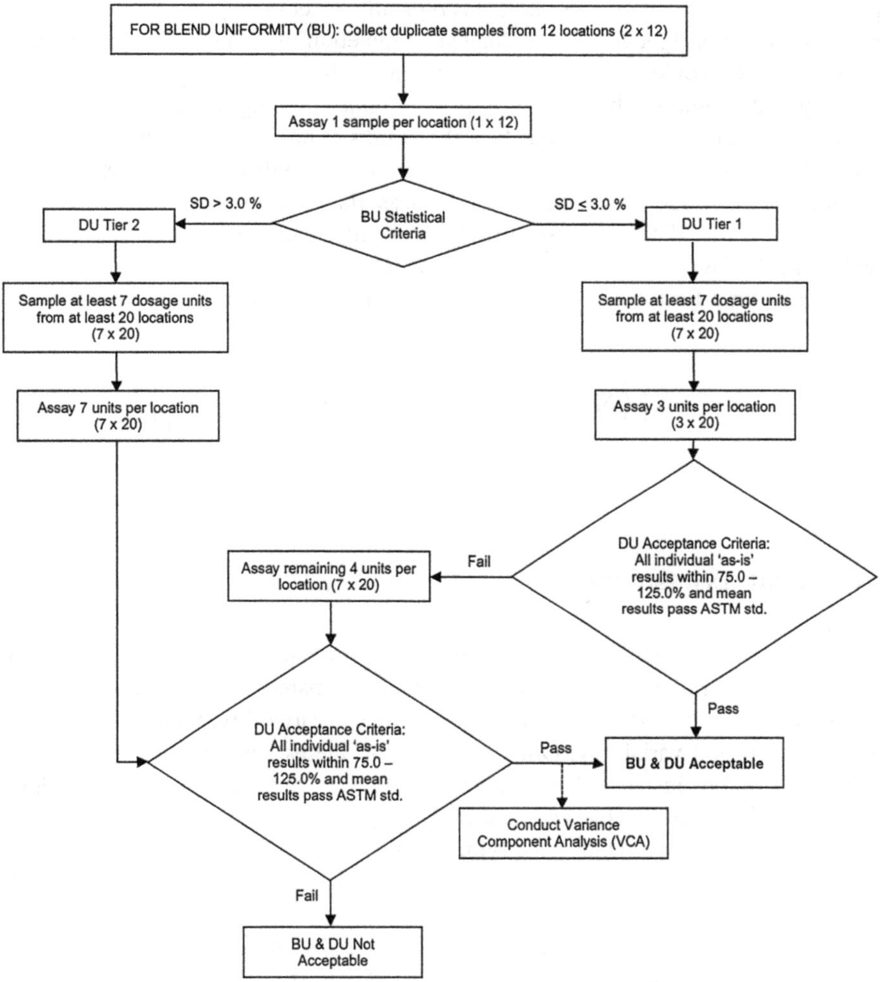

Fig. 6.3 BU and CU sampling and testing plan

Post PPQ Stage 3a Assessment

Stage 3a is the initial assessment post-new product launch that utilizes a substantial body of data for statistical evaluation to gain a deeper product understanding. A Stage 3a assessment [6] utilizes data from all process validation stages. Stage 3a assessment is thus pivotal in understanding and managing product variability since the product control strategies are based on early estimates of process capability at the initial commercial launch time. The assessment of the predetermined number of batches, therefore, augments the initial process knowledge gained from quality by design (QbD) based product development and verification (PPQ) stages. A Stage 3a protocol with the requirements defined is typically generated upon completion of the process validation lifecycle Stage 2b. The completion of a Stage 3a report demonstrates the organization's compliance with establishing an enhanced product control strategy and attaining a high level of product understanding and quality. To determine the number of Stage 3a batches, the results from Stage 2b PPQ batches are utilized. For example, USP <905> dosage uniformity indicates the computation of an acceptance value (AV) that must be less than 15 to meet the L1 criteria. Preexisting batch data was used to determine the inter-batch variability (SB-B), and Stage 2b data was used to determine the intra-batch variability (S0). The upper confidence limit for the estimated number of Stage 3a batches is derived from:

$$S^{\mathrm{hi}}_{(N_{\mathrm{B}})} = \sqrt{\left\{ S_{\mathrm{B\text{-}B}}{}^{*} \sqrt{\frac{(N_{\mathrm{B}}-1)}{\chi^{2}{}_{\left(\frac{\alpha}{2}, N_{\mathrm{B}}-1\right)}}} \right\}^{2} + \left\{ S_{0}{}^{*} \sqrt{\frac{(N_{0}-1)}{\chi^{2}{}_{\left(\frac{\alpha}{2}, N_{0}-1\right)}}} \right\}^{2}}$$

Statistical Assessment

Inherent process variability (IPV) measures batch-to-batch variability, while analytical (method) variability measures the variability of material within the same batch. As such, estimates of ${\sigma^2}_a$ measure inherent process variability, while ${\sigma^2}_e$ measures analytical method variability. Other measures of interest can be obtained from the above model. For instance, the ratio of these two variance components provides a standardized measure of the variance of the population group means. At the same time, the intra-class correlation measures the proportion of the total variance due to the process. Estimates for these values can be obtained from the ANOVA data provided (Table 6.1).

Other estimators are available, in particular for unbalanced data where a different number of measurements are taken per batch. The restricted maximum likelihood (REML) estimator is a viable alternative available in most statistical software packages. This model can be fit to situations in which the batch effect is considered random, and each batch has n samples. For example, 20 batches might be considered

Table 6.1 Estimates

Source of variation	Degrees of freedom	Sum of squares	Mean square	Expected mean square
Between groups (Process)	$r-1$	$\mathrm{SSA} = n\sum_{i=1}^{r}(\overline{y_{i.}} - \overline{y_{..}})^2$	$\mathrm{MSA} = \frac{\mathrm{SSA}}{(k-1)}$	$n\sigma_a^2 + \sigma_e^2$
Within groups (Analytical)	$r(n-1)$	$\mathrm{SSE} = \sum_{i=1}^{r}\sum_{j=1}^{n}(y_{ij} - \overline{y_{i.}})^2$	$\mathrm{MSE} = \frac{\mathrm{SSE}}{k(n-1)}$	σ_e^2
Total	$rn-1$	$\mathrm{SST} = \sum_{i=1}^{r}\sum_{j=1}^{n}(y_{ij} - \overline{y_{..}})^2$		

to be a random sample from a larger pool of batches for a specific product. For each of these 20 batches, a random sample of 10 samples would be taken to measure finished product dosage uniformity. The variability in the mean finished product dosage uniformity between the batches, σ2a, would yield an estimate of the inherent process variability, while ρa would provide an estimate of the proportion of variability due to the process. Confidence intervals can be constructed for these estimates and are available in common statistical software packages. The estimated IPV, as well as the ratio of total variance due to the process, can be used during Stage 3 batch monitoring to focus efforts on process improvement. As more information is gathered for a product, a rise in the IPV itself, or a rise in the proportion of total variance due to the process (ρa), could indicate the need to investigate possible process improvements. On the other hand, a decrease in IPV or the decline in ρa would indicate that the process is improving. In order to obtain a picture of how well the process is performing overall for a specific product, a comparison with other products employing the same process can be made by generating a benchmark. The PaCS index [7] provides an indication of a current product's process performance in comparison to other similar products. To derive the PaCS index, a representative set of other products generated with the same process would be chosen. For each of these chosen products, the IPV would then be calculated as above. The PaCS index could then be calculated using the following equation:

$$\mathrm{PaCS} = \mathrm{IPVp}/\mathrm{IPVb}$$

where IPVb is the benchmark inherent process variability and IPVp is the inherent process variability for the product under consideration. IPVb is the median process variability of the selected products with processes similar to the current product.

A PaCS index greater than 1 indicates that the process variability is high, while a PaCS less than 1 indicates that process variability is low compared to the benchmark. Therefore, a PaCS value that is less than 1 is preferred. Because the distribution of the PaCS index is not analytically derivable, confidence intervals can be estimated using Monte Carlo simulation. The PaCS index, together with IPV values and the other derived statistics, provides a platform upon which further decision-making can take place. For instance, high PaCS values could indicate that the process for a specific product is not performing as expected. Estimation of inherent process variability (IPVP) allows for determining a PaCS index for the product and helps in understanding the contribution to variability that comes from the manufacturing process and the analytical method used. In addition, PaCS is a metric developed in relation to the manufacturing process at a particular production site. The index can effectively determine continuous improvement projects at the site or for site transfer initiatives. PaCS provides a tangible quantitative robustness figure for various supply chain decision-making scenarios. The index can be a component of a periodic process performance review by senior management as recommended by ICH Q10. In addition, APV and the PaCS Index may be used to decide such things as who should be primarily responsible for a specific continuous improvement project (i.e., whether process, analytical, or a combination). This is often a point of contention. It

Table 6.2 Example

Product	Average DU	$\hat{S}^2_\alpha$	$\hat{S}^2_e$	$\hat{S}^2_y$	$p^\wedge_\alpha$	$p^\wedge_e$
Product 1	99.7 (98.7,100.6)	1.61 (0.71,5.60)	1.02 (0.77,1.39)	2.63 (1.70,6.63)	0.61 (0.39,0.85)	0.39 (0.15,0.61)
Product 2	99.1 (98.3,100.0)	0.97 (0.20,4.28)	4.50 (3.43,6.17)	5.47 (4.25,9.11)	0.18 (0.04,0.49)	0.82 (0.51,0.96)
Product 3	100.5 (99.6,101.4)	1.25 (0.46,4.70)	2.33 (1.77,3.19)	3.57 (2.65,7.11)	0.35 (0.15,0.67)	0.65 (0.33,0.85)
Product 4	98.8 (97.9,99.6)	0.96 (0.18,4.30)	4.72 (3.60,6.48)	5.68 (4.42,9.38)	0.17 (0.03,0.48)	0.83 (0.52,0.97)
Product 5	100.0 (99.2,100.8)	1.17 (0.48,4.18)	1.28 (0.98,1.76)	2.45 (1.71,5.50)	0.48 (0.26,0.77)	0.52 (0.23,0.74)
Product 6	99.5 (99.1,99.8)	0.18 (0.06,0.73)	0.52 (0.39,0.71)	0.70 (0.53,1.28)	0.26 (0.09,0.59)	0.74 (0.41,0.91)
Product 7	99.2 (98.7,99.6)	0.39 (0.15,1.44)	0.63 (0.48,0.87)	1.02 (0.75,2.10)	0.38 (0.18,0.70)	0.62 (0.30,0.82)
Product 8	101.2 (100.8,101.7)	0.23 (0.00,1.19)	1.84 (1.40,2.52)	2.07 (1.62,3.21)	0.11 (0.00,0.40)	0.89 (0.60,1.00)
Product 9	98.2 (97.9,98.5)	0.07 (0.00,0.49)	1.11 (0.84,1.52)	1.18 (0.92,1.74)	0.06 (0.00,0.31)	0.94 (0.69,1.00)
Product 10	98.9 (98.4,99.3)	0.23 (0.03,1.07)	1.35 (1.03,1.85)	1.58 (1.23,2.54)	0.14 (0.02,0.45)	0.86 (0.55,0.98)
Product 11	99.5 (99.1,100.0)	0.18 (0.00,1.26)	2.89 (2.20,3.96)	3.06 (2.40,4.52)	0.06 (0.00,0.31)	0.94 (0.69,1.00)

could also be used to determine which site has the best PaCS index with respect to a product. This factor will be considered when deciding for or against site product volume increases. In summary, the PaCS can provide valuable insight to decision-makers and help to drive continuous quality improvement programs in biopharmaceutical and pharmaceutical development as well as manufacturing (Table 6.2).

Given that IPVp is 0.09 and IPVb of 0.39, the PaCS index is calculated as 0.09/0.39 = 0.23, indicating that the process variability for the FCT product is low compared to the benchmark. A product-specific Process Capability and Quality Dashboard (PCQd) is a critical component of Stage 3a assessment in projecting product robustness. The dashboard addresses the elements in the FDA's Guidance: Request for Quality Matrix, where the agency suggests optional metrics as evidence of manufacturing robustness and a commitment to quality. Data reported indicates low risk and may merit a reduction of site inspection frequency. Each processing stage can be evaluated against a predetermined process capability target to provide an overall product performance synopsis (Table 6.3).

A trained statistician is responsible for selecting statistical tools employed in this assessment, i.e., evaluating process stability and capability. PCQd targets may be set before Stage 3a initiation based on accepted process performance indices (e.g., Pp, Ppk) and stringent control limits. Pp formula considers the extent of variation given by standard deviation and an acceptable range allowed by specified limits despite the mean and hence is appropriate for in-process CQAs such as Sieve analysis. For IP

Table 6.3 Targets

	Process capability targets	Product performance
Compaction		
Sieve results	Pp > 1	2.2
Compression		
IP hardness	Ppk > 1	1.3
IP weight variation	Ppk > 1	1.7
Press speed	All batches are to be run at max. validated speed	Meets
% yield	Meets alert limit as per SOP	Meets
Encapsulation		
Weights	Ppk > 1	1.3
Speed	All batches are to be run at max. validated speed	Meets
% yield	Meets alert limit as per SOP	Meets
FP CQA		
FP disso	Pa > 99.9%	Meets
FP assay	Pa > 99.9%	Meets
FP CU	AV < 10	Meets
Compliance		
# of process deviations	Zero	Meets

CQA, the sample is intended to meet the specification requirement at predefined intervals through the manufacturing process. The Ppk estimate can diagnose decentralization problems aside from the process variation. As such, it applies to hardness and weight variation where meeting the target specification is the objective.

Probability of acceptance (Pa) is applicable for CQAs having stage-wise acceptance criteria, such as dissolution, where traditional process capability measures are inadequate. Some of the benefits of implementing PCQd include enabling proactive risk mitigation activities, empowering management with product performance oversight, improving supply chain predictability and manufacturing reliability, encouraging the implementation of emerging technology to reduce variability, and enabling regulators (e.g., US FDA) in developing a risk-based site inspection schedule. Organizations may also utilize PCQd as a transfer criterion between product development and commercial operations responsible for the commercial lifecycle management of the product.

ICH Q10 Pharmaceutical Quality System guidance states that process performance and product quality monitoring systems should provide tools for measurement and analysis of critical material attributes (CMAs), critical process parameters (CPPs), and critical quality attributes (CQAs) identified in the control strategy. An enhanced product control strategy is finalized based on quality by design (QbD) product development data, qualification results, and results from additional Stage 3a commercial batches. FDA recognizes the importance of utilizing post-launch learning that may be used to enhance product control strategy. CMAs for different lots of raw materials used in the manufacturing of Stage 3a batches may be assessed. Even though raw material lots meet vendor and in-house specification limits, it is important to ensure that trends observed in any identified CMAs are not adversely impacting finished product quality attributes such as dissolution, content uniformity, and assay. Additional process or specifications controls may be required for drug product and/or drug substance manufacturing. CPPs evaluated at Stage 3a provide insight into any process drifts. Statistical process control charts may be used for each CPP to evaluate the process parameter variability. Any observed trends may need continuous improvement actions to be taken. CPPs are defined in Stage 1 based on QbD-based product development. The extent of the CPPs' impact on CQAs is analyzed in case high variability is observed for CQAs, e.g., parameters such as precompression force and main compression force on dissolution. Statistical models may be built to understand any new relationships that may surface in this further assessment and serve as a critical component in developing process understanding and control strategy.

In-process quality attributes at each manufacturing stage, such as compaction, compression, and coating, and finished product quality attributes, such as assay, dissolution, and dosage uniformity, are evaluated. A process is determined to be in a state of statistical control when it produces products that are stable and predictable over time, despite a certain amount of variability in the CQAs. Statistical process control charts are the most common tool in process monitoring. Process performance analysis (Ppk) and probability of acceptance analysis (Pa) are also used to provide quantifiable evaluations and predictions of product performance. Analysis of results from Stage 3a batches can then be used to demonstrate the probability of future batches meeting CQA specifications. A risk-based scientific assessment, along with the design of experiment (DoE) studies, allows a control strategy to be established in Stage 1, subsequently verified during Stage 2 and further enhanced during Stage 3a (Table 6.4).

Defining a Stage 3b monitoring plan is part of the enhanced 3a control strategy for product lifecycle management. Further understanding of sources of variability and their impact on downstream processes, in-process materials, and drug product quality provides an opportunity to shift controls upstream and minimize the need for end-product testing. Alternative approaches to meeting quality commitments (e.g., replacement of blend uniformity with stratified dosage uniformity or process analytical technologies such as NIR) are justified with the availability of additional data. The following assessments are typically conducted as part of the Stage 3a review (Figs. 6.4, 6.5, and 6.6).

The Stage 3a summary enables the initiation of continuous improvement projects for further product enhancement and optimization.

Stage 3b Routine Monitoring

Stage 3b is routine continued process verification and trending of CMA's, CPP's and IP and FP CQAs within established alert limits. The Stage 3b CPV program allows us to comply with the lifecycle guidance and avoid regulatory risks. Stage 3b ensures

Table 6.4 Parameter control strategy

			Post stage 1			Post stage 3a			Granule PSD	
Compaction process parameter	Critical	Early-stage 1	Min	Max	Target	Min	Max	Target	Mesh	Spec.
Force (KN/cm^2)	Yes	4-12	6	10	8	6	10	8	20	NLT 25
Roller gap (mm)	Yes	2.0–4.0	2.0	4.0	2.5	2.0	3.0	2.5	40+60	25–45
Roller speed (rpm)	Yes	2–12	5	10	8	6	10	8	80+100	25–45
Gran speed (rpm)	No								Fines	NMT 30

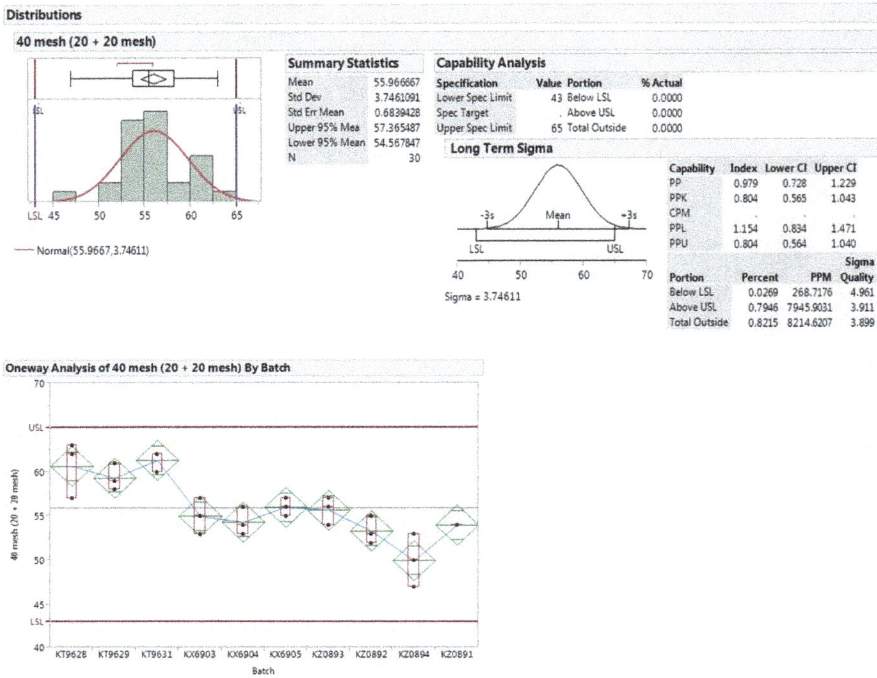

Fig. 6.4 Trending of Sieve profile

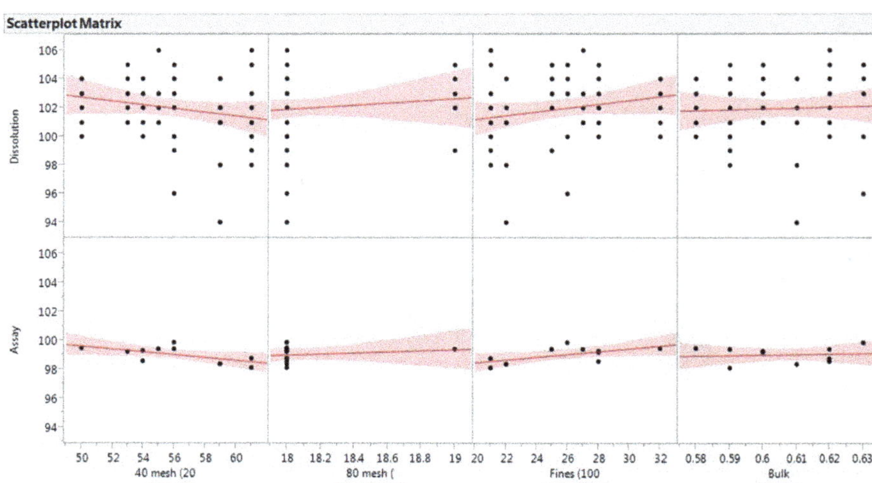

Fig. 6.5 Determining compaction and FP CQA correlation

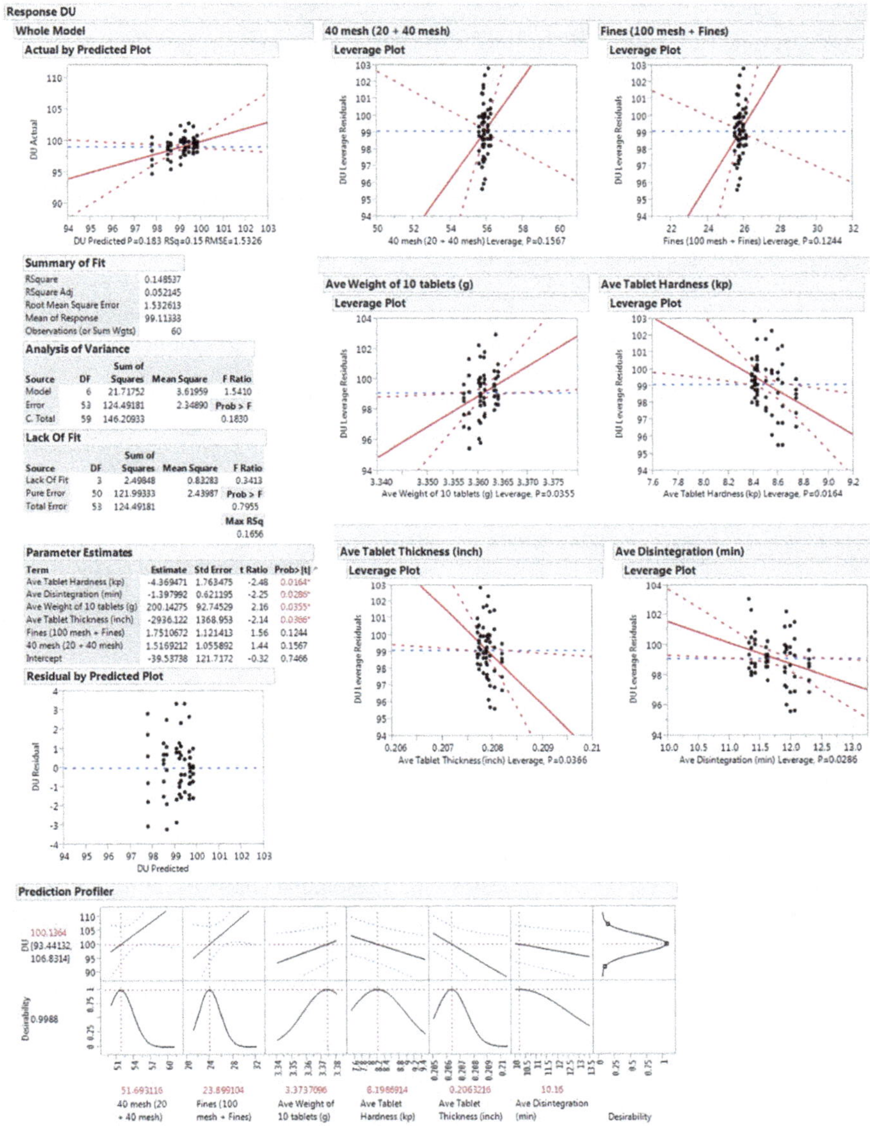

Fig. 6.6 Regression analysis for IP CQAs and DU

that during routine production, the process remains in a state of control. A system or systems for detecting unplanned departures from the process as designed is essential to accomplish this goal. Collection and evaluation of process performance data will allow the detection of process drift. The information collected should verify that the quality attributes are being appropriately controlled throughout the process. If properly carried out, these efforts can identify variability in the process and/or signal potential process improvements. Statistical process control (SPC) tools primarily

Table 6.5 Rules

Rule	Description	Possible concern
Rule 1	One point more than three standard deviations from mean	Indicates a statistically anomalous event
Rule 2	Nine sequential points on the same side of the mean	Potential prolonged bias
Rule 3	Six sequential points continually decreasing or increasing	A potential trend
Rule 4	Fourteen sequential points alternate (oscillate) in the direction	Potential multiple underlying processes
Rule 5	Two (or three) out of the three points in a row are more than 2 standard deviations from the mean in the same direction	There is a medium tendency for samples to be mediumly out of control
Rule 6	Four (or five) out of five points in a row are more than 1 standard deviation from the mean in the same direction	There is strong tendency for samples to be slightly out of control
Rule 7	Fifteen points in a row are all within 1 standard deviation of the mean on either side of the mean	Greater variation expected
Rule 8	Eight points in a row exist with none within 1 standard deviation of the mean and the points are in both directions of the mean	Moving above to below and missing the first SD band is rarely random

used in Stage 3b were initially developed by Walter Shewhart and gained popularity following W. Edwards Deming's implementation in the automobile industry. SPC trend limits, coupled with control chart rules (e.g., Western Electric or Nelson Rules) [8], alert to potential non-random events or deviations. There is a multitude of SPC charting rules that may be useful to identify potentially statistically anomalous events. These include (Table 6.5):

Triggering one of such rules indicates with reasonable statistical confidence that something may have changed within the process that may impact the product's robustness and control. Signals should trigger a response reaction based on the risk. Stage 3b CPV efforts can identify variability in the process and/or signal potential process improvements as a process is likely to encounter sources of variation. Statistical signals, however, need not be classified as deviations and investigated immediately. A controlled manufacturing process is expected to see acceptable "process shift" signals or "yellow flags," and not all signals are created equally. The reaction strategy should therefore depend on the influence on patient impact CQAs.

Stage 3b (CPV) covers many annual product quality review components. Continuous monitoring and analysis of process data through CPV, within established limits, allows organizations to implement an effective product control strategy. CPV signals [9] can indicate issues related to quality attributes, material attributes, process parameters, and performance indicators and can vary in their significance level. It is important to categorize these signals based on risk and take appropriate action based on the signal received. When a significant amount of data is available, an action plan

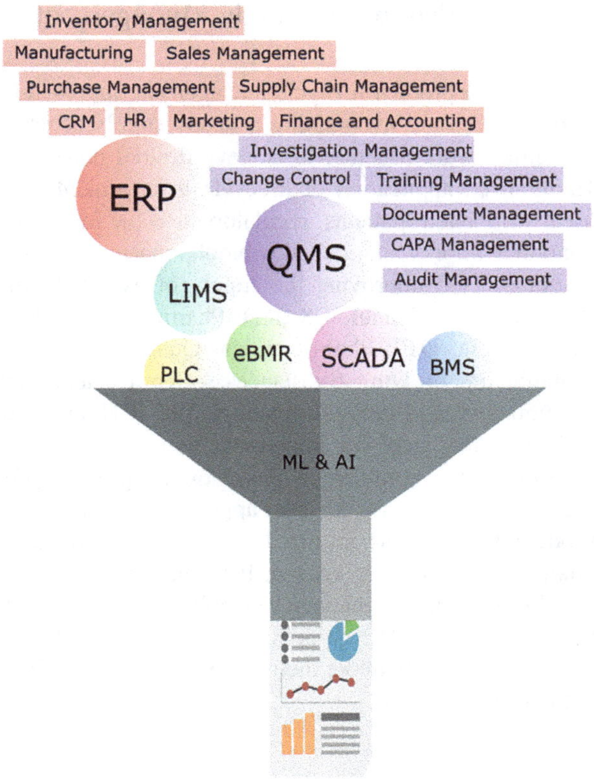

Fig. 6.7 Integrating GxP and business systems for knowledge management

can be implemented. The chosen course of action is determined by the practical relevance and statistical strength of the signal.

To comply with regulations, integration of data from material attributes, process parameters, and quality attributes is necessary. This is typically achieved with electronic systems such as an enterprise resource planning (ERP) system, laboratory information management system (LIMS), quality management system (QMS), manufacturing equipment, and other electronic data sources (Fig. 6.7). Easy and quick access to data from the first stage of development and the integration of commonly used statistical assessment tools and risk assessment software are also crucial for implementing a smooth process validation lifecycle management system that meets regulatory expectations. The process validation lifecycle management system (PVLMS) should have all data from stages 1, 2, and 3 in a single, easily accessible platform. Data from all stages makes it easier to manage knowledge and use data through effective search query functions. Integrated statistical analysis tools, risk assessment tools, and online tools for monitoring progress in stage 3 (CPV) can greatly benefit any organization. Effective use of data from stage 3 (CPV) is a vital business requirement in the pharmaceutical industry. It can lead to reduced process failure rates, early identification of process losses, and even the reduction of developmental studies using similar process data.

ICHQ12: Product Lifecycle Management

An effective product lifecycle management approach [10] is crucial for continuous improvement and maintaining a steady supply of drug products. Lifecycle-related guidelines work together to achieve desired outcomes as expected by regulators. Harmonized application of lifecycle management at the regulatory and technical level will benefit patients, regulators, and the industry. It promotes innovation and continual improvement in the biopharmaceutical industry, strengthening quality assurance, and improving the supply of essential drug products to patients. The ICH quality guidelines, Q8 (R2): Pharmaceutical Development, Q9: Quality Risk Management, Q10: Pharmaceutical Quality System, and Q11: Development and Manufacture of Drug Substances, offer a science and risk-based approach for development and decision-making. The ICH Q12 guideline addresses technical and regulatory gaps for implementation to fully realize the guidelines. The Q12 focuses on the commercial phase of the product lifecycle and requires flexible regulatory approaches for post-approval CMC changes. The Q12 is dependent on product and process knowledge, risk management principles, and an effective pharmaceutical quality system. It facilitates post-approval CMC changes in a predictable and efficient manner. The effective implementation of the tools and enablers described in the Q12 guideline enhances the industry's ability to manage CMC changes effectively under their Pharmaceutical Quality System (PQS) with minimal regulatory oversight.

Established Conditions and PACMP

A control strategy is a plan for ensuring consistent production of a product of required quality based on the current understanding of the product and process (ICH Q8(R2)). It includes controls for drug substance and drug product, materials and components, facility and equipment operating conditions, in-process controls, finished product specifications, and the methods and frequency of monitoring and control (ICH Q10). Established conditions (ECs) are legally binding information necessary to assure product quality, therefore, any changes to ECs require submission to the regulatory authority.

ECs for a manufacturing process are defined based on product and process understanding, considering all relevant elements of the control strategy. In addition to the unit operation and sequence of steps, ECs proposed and justified in the manufacturing process description should be inputs (e.g., process parameters, material attributes) and outputs (including in-process controls) necessary to assure product quality. Process parameters that must be controlled to ensure a product of required quality will be produced are considered ECs. These ECs are identified through an initial risk assessment and application of knowledge gained from executed studies, prior knowledge, and a criticality assessment that determines the level

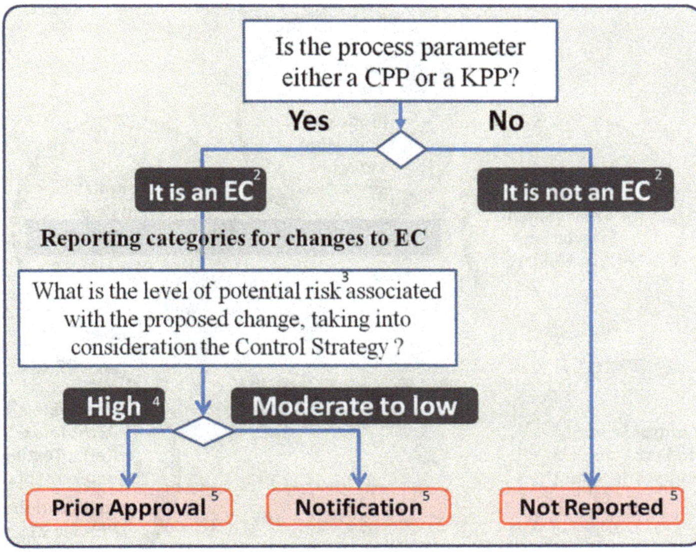

Fig. 6.8 Established conditions

of impact a process parameter could have on product quality (Fig. 6.8). The criticality assessment account for the severity of harm and whether the ranges studied sufficiently account for the expected variability in the EC. CPPs and other process parameters where an impact on product quality cannot be reasonably excluded are identified as ECs. Once ECs are identified, an updated assessment of the potential risk to product quality associated with changing the EC, taking into account the overall control strategy, informs the reporting category for the EC. The assessment of potential risk is derived from risk management activities. The output of the risk assessment includes changes to manufacturing process ECs that range from high to low risk to product quality. The reporting category is defined based on the level of risk. A justification of the potential risk for changing ECs and corresponding reporting categories is submitted.

The industry has been heavily investing in the implementation of quality by design (QbD) based product development (ICH Q8, Q11), risk management (ICH Q9), and pharmaceutical quality systems (ICH Q10) over the past decade. Before ICH Q12 (Technical and Regulatory Considerations for Pharmaceutical Product Lifecycle Management), there were no mechanisms to reward organizations implementing risk-based and data-driven development approaches. ICH Q12 closes this gap where organizations can fully realize the benefit: of implementing lifecycle changes with ease and significantly reduced regulatory burden. This is a win-win situation for the industry, regulators, and patients. A well-planned product lifecycle change management (across various markets) is key to ensuring an uninterrupted supply of pharmaceutical products. The variation/change guidelines and regulatory waits were essentially making it difficult to implement continuous improvement

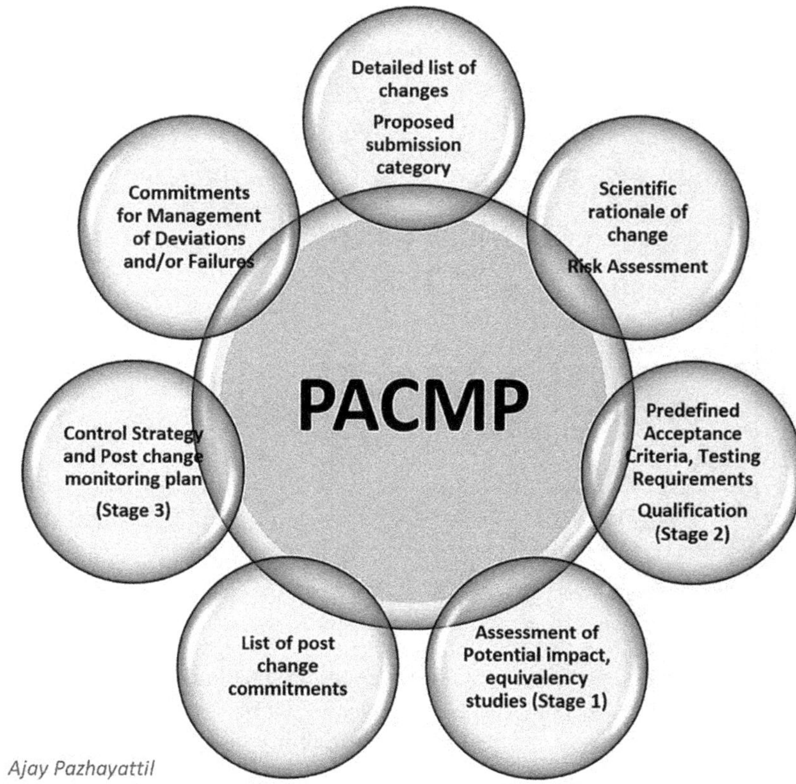

Fig. 6.9 Post Approval Change Management Protocol (PACMP) components

changes. The current regulatory shift makes investment in QbD and implementation of lifecycle concepts, including process validation continued/ongoing process verification stage, fruitful.

The ICH Q12 provides a tool: post approval change management protocol (PACMP) [11] to make predetermined changes without impacting your product supply. PACMP defines acceptance criteria and requirements for the typical lifecycle changes such as process scale-up, API, and equipment (Fig. 6.9). Once submitted, PACMP saves considerable time and resources. Effective application of PACMP is a great organizational strategy for future product lifecycle change management.

References

1. ISPE (2019) Baseline guide vol 5: commissioning & qualification. International Society for Pharmaceutical Engineering
2. ASTM (2020) ASTM 2500 standard guide for specification, design, and verification of pharmaceutical and biopharmaceutical manufacturing systems and equipment. American Society for Testing and Materials
3. ISPE (2019) Good practice guide: process validation. International Society for Pharmaceutical Engineering
4. Pazhayattil A, Alsmeyer D, Chen S, Hye M, Ingram M, Sanghvi P (2015) Stage 2 process performance qualification (PPQ): a scientific approach to determine the number of PPQ batches. AAPS PharmSciTech 17(4):829–833
5. Sayeed-Desta N, Pazhayattil AB, Collins J, Doshi C (2018) A science and risk-based pragmatic methodology for blend and content uniformity assessment. AAPS PharmSciTech 19(3): 1483–1492
6. Sayeed-Desta N, Pazhayattil AB, Collins J, Chen S, Ingram M, Spes J (2017) Assessment methodology for process validation lifecycle stage 3A. AAPS PharmSciTech 18(5):1881–1886
7. Collins J, Sayeed-Desta N, Pazhayattil A, Doshi C (2017) A continuous improvement metric for pharmaceutical manufacturing. Pharmaceutical Technology
8. Montgomery DC (2005) Introduction to statistical quality control, 5th edn. Wiley, Hoboken
9. Pazhayattil A (2020) Continued process verification: reacting to data signals. PDA Letter
10. RAPS (2021) Fundamentals of pharmaceutical and biologics regulations, 4th edn. Regulatory Affairs Professionals Society
11. Pazhayattil AB, Sayeed-Desta N, Iyer V (2017) ICH Q12 post approval change management protocol: advantages for consumers, regulators and industry. Regulatory Affairs Professionals Society

Chapter 7
Continuous Improvement Case Study: Transforming Legacy Products

Abstract This chapter describes a systematic approach to the continuous improvement of legacy products, using principles of pharmaceutical manufacturing science and statistical tools. The goal is to minimize variability, understand the impact of processing and analytical parameters on critical quality attributes, and develop targeted improvement plans. The methodology is in line with the 2011 US FDA Process Validation Guidance for Industry and allows for patient risk mitigation by generating data that is otherwise unavailable during the lifecycle of a legacy product. The discussion focuses on the heightened sampling and testing applied while monitoring legacy products. This approach provides an opportunity to determine the robustness of the legacy process. Additionally, it enables organizations to develop a greater understanding of the legacy product, infer the sources of variability, and control it in a manner commensurate with the risk it represents to the process and product. The insights gained can be used to support science- and risk-based decisions for the formulation and manufacturing processes of similar products.

Keywords Continuous Improvement · Legacy Products · Process Variability · Continued Process Verification · Process Knowledge · Data Analysis · Manufacturing Science

Introduction

This chapter presents a structured approach for making science- and risk-based decisions for continuous process improvement with increased testing and sampling of legacy products. The approaches utilize principles of pharmaceutical manufacturing science and statistical tools to determine the robustness and variability of existing powder for oral suspension products. The goal is to minimize sources of variability to understand the impact of processing and analytical parameters on critical quality attributes and develop targeted continuous improvement plans. This methodology aligns with the 2011 US FDA Process Validation Guidance for Industry. It helps to understand sources of variability, the level of variation, and the impact of variability on critical quality attributes of oral powder for suspension products.

A. B. Pazhayattil et al., *Technology Transfer*, AAPS Introductions in the Pharmaceutical Sciences 10, https://doi.org/10.1007/978-3-031-32192-4_7

The discussion reviews the heightened sampling and testing applied for powder for the legacy product (powder for oral suspension) during Stage 3. Stage 3 monitoring of legacy processes allows for the observation and/or collection of statistically relevant data and the corresponding sets of parameters and attributes at key processing stages to apprehend the impact of variability (inter/intra batch variability). The data can be utilized to determine the current state of control of the manufacturing process. Heightened Sage 3 sampling and testing of legacy products present an opportunity to determine the robustness of the legacy process and determine the need for continuous improvement. It has to be emphasized that processes inherently determined to be variable through risk assessment and historical data need to be on a remediation plan (Stage 1). The products are to be discontinued from the market when high variability is determined based on the Stage 3 data assessment. The sampling and testing approach discussed enables patient risk mitigation since the newly generated data is otherwise unavailable during the lifecycle of a legacy product that continues to successfully meet the approved quality standards.

Method

The Stage 3 batches of the powder for oral suspension products shall follow the approved commercial master batch record and to be manufactured, filled, and tested under normal conditions (utility systems (e.g., air handling), material, environment, manufacturing procedures, and approved test methods) by trained personnel who routinely perform each step of unit operations (Fig. 7.1). Heightened samples of a statistically relevant number are collected in addition to the routine commercial sampling and testing. The execution of data-driven Stage 3 heightened monitoring through a protocol that enables continual enhancement of the product control strategy. It closes any process characterization gaps that were not addressed during non-quality by design (QbD) legacy process development. The initial heightened Stage 3 sampling and testing allows for the development of a scientifically sound ongoing Stage 3 program as recommended by US FDA's Process Validation Guidance for Industry and to meet the QbD standards according to ICH Q8 (R2): Pharmaceutical Development.

The material attributes, processing parameters, and quality attributes are critically evaluated during the initial Stage 3 assessment of the legacy product manufacturing process. Being a legacy product, and in the absence of a QbD (ICH Q8(R2) dated 2009) based product development report, a combination of documents such as initial product development report, process validation reports, batch manufacturing and packaging documents, specifications, test methods, and initial and post-approval submission documents are assessed to determine the existing product control strategy. The critical quality attributes (CQA), critical process parameters (CPP), in-process and finished product specifications, and initial Stage 3 statistical criteria are defined in the protocol. The adequacy of the submitted specifications is also

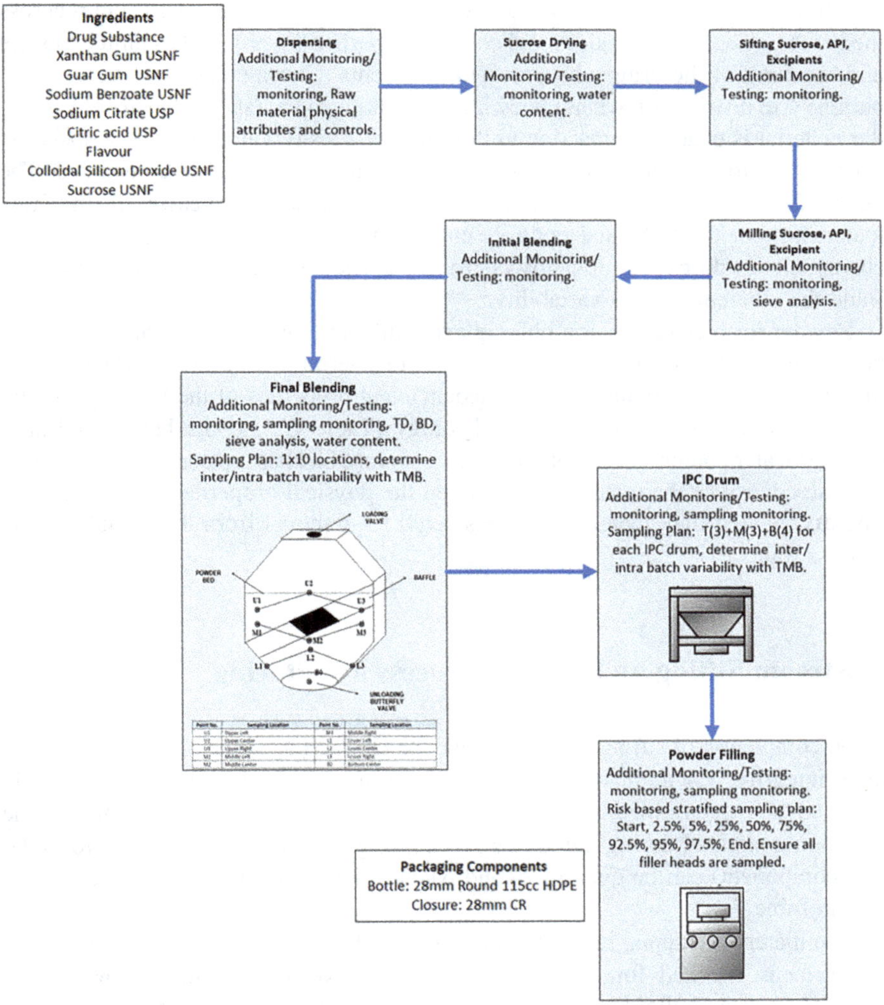

Fig. 7.1 PfOS manufacturing process steps example

reviewed at this juncture. Statistical assessments are conducted to diagnose the underlying patterns for the purpose of determining the need for continuous improvement and for gathering additional process knowledge at scale.

Material Attributes and Formulation Monitoring

The raw material batches can be selected from recently manufactured batches and reviewed to ensure that they did not have deviations and out-of-specification (OOS) results. In addition, a single excipient batch and a single API batch may be used in the initial Stage 3 batches. This is to reduce any impact of material attributes across the batches. There can be additional controls as part of the protocol to limit exposure and handling of the drug substance and excipients. Consideration can be given to particle size distribution when reviewing the bulking agent (such as sucrose) batch if the material is coarser (impacting uniformity). Viscosity values for the suspending agent in the formulation, such as xanthan gum and guar gum, are considered. The deliberate use of similar grade material for Stage 3 batches, the preassessment of raw materials, and the tightened controls ensure that the API and excipient physical characteristics do not significantly contribute to the variability of CQAs, to understand the inherent process variability.

Powder for oral suspension bulking agent/diluent (e.g., sucrose) can be sampled before and after drying (if part of the process) to determine the effect of water content in downstream processing since physicochemical properties of the largest formulation component may affect powder rheology (Table 7.1). If the before and after (drying) water content results of the Stage 3 sucrose batches are constant, it confirms that the control of the material ensures that the physical properties are maintained. The drying step may therefore be considered for removal from the manufacturing process train.

Upstream Sifting and Milling Process Monitoring

Upon completion of sifting, the sieves may be checked to ensure that they maintain their integrity. Each quantity of sifted material collected may be individually weighed and determined to see the existence of batch-to-batch variability. The retention generated during the sifting process, e.g., bulking agent (largest formulation component) can be quantified, and batch-to-batch consistency of retains measured (Table 7.2).

The materials trapped inside the milling chamber are also ideal to be quantified and screens checked for blinding post milling process. Milling of low-density material such as colloidal silicon dioxide USP-NF can result in dusting (and in process loss) during milling operation and ultimately affect downstream powder blend flow properties. Hence, such focused observations are critical during Stage 3. The milling step in the powder for suspension formulation is sometimes applied to enhance the cohesiveness of drug substance and the bulking agent such as sucrose. This strategy is expected to improve the mixing efficiency and minimize the likelihood of potential segregation. However, it should be noted that the formulation needs to be optimized to reduce the gap between the particle size distribution (PSD) of the API compared to the bulking agent. As such, for the purpose of legacy product knowledge enhancement, an unmilled mix may be prepared to utilize the same API and excipient batches and compare. The unmilled mix, the milled mix from the Stage 3 batches, and a placebo mix sample can be analyzed (Fig. 7.2) to document the impact of milling on particle size and crystal structure solid-state characteristics. X-ray diffraction (XRD) peaks can define the characteristics of the mix, while the

Table 7.1 An example of additional stage 3 CMA and formulation monitoring elements for a legacy PfOS

Additional stage 3 Process monitoring plan	Additional stage 3 Sampling and testing plan	Rationale for 3 assessment
Unit operation: raw material dispensing [dispensing booth]		
1. Identify physical attributes of the Raw Material (RM) batches to be used 2. Use one RM batch only 3. Reviewed gum viscosity results 4. Verify use of sealed container 5. Monitor transfer of raw materials and in process storage 6. Observe for any static in the polyethylene bag/container 7. Verify material container and transfer container integrity 8. Observe dispensing/ weighing activity	1. Dispense separately API for water content before use for API water content (KF) 2. Dispense separately unmilled mix for XRD 3. Dispense separately unmilled mix for DSC 4. Dispense separately unmilled mix for PSD	1. To determining potential source of variability 2. Confirm reduced variability by selecting one RM batch 3. Enhancing legacy process knowledge
Unit operation: sucrose drying [fluid bed dryer]		
1. Verify FBD functionality 2. Ensure fixed FBD parameters are being used across monitoring batches 3. Document raking performed 4. Record container seal pressure at start 5. Record filter bag knob air pressure 6. Verify continuous drying without stoppage 7. Verify if Sucrose is entrapped in finger bag 8. Ensure process continues to next stage without delay 9. Monitor transfer conditions of Sucrose	1. Sample Sucrose for water content (KF) before drying 2. Sample Sucrose for water content (KF) after drying	1. Determining potential source of variability 2. Enhancing legacy process knowledge

placebo will be devoid of interference. The XRD of milled and unmilled batches can indicate the impact of milling on the solid-state properties of the mix. Differential scanning calorimetry (DSC) can exhibit characteristic transitions for milled material. The DSC data, therefore, can support the legacy formulation strategy of enhancing cohesiveness by including the milling step. Finally, the thermograms for all the Stage 3 milled mixes can be compared, thus displaying that the milling process may not contribute to CQA variability.

Table 7.2 An example of additional stage 3 upstream operational parameter monitoring

Additional stage 3 Process monitoring plan	Additional stage 3 Sampling and testing plan	Rationale for 3 assessment
Unit operation: sifting [mechanical sifter]		
1. Verify equipment functionality 2. Ensure fixed sifting parameters 3. Use the same API polybag for co-sifted material to minimize API loss 4. Ensure earthing to sifter 5. Record Sucrose retained quantity in sifter 6. Ensure process continues to next stage without delay 7. Record weight of the sifted materials 8. Monitor transfer conditions	Not applicable	1. Determining potential source of variability 2. Enhancing legacy process knowledge
Unit operation: milling [comminuting mill]		
1. Verify equipment functionality 2. Ensure fixed milling parameters across batches 3. Monitor order of addition 4. Check for static in polybag 5. Ensure earthing 6. Verify uniform feed 7. Verify for melting or other observations 8. Monitor unloading of milled material into IPC 9. Open the milling chamber to ensure no entrapped material 10. Ensure process continues to next stage without delay 11. Monitor transfer conditions and process	1. Sample for XRD 2. Sample for DSC 3. Sample for PSD	1. Characterization of key process step involving milling of sucrose with API & excipients 2. Determining potential source of variability 3. Enhancing legacy process knowledge

Mixing Process Parameter Monitoring

Distance measurements can be taken from the top of the blender to the surface of the blend with a calibrated laser device after loading the materials for initial blending, after the addition of the remaining material, and after the completion of final blending. 3D surface plots (Fig. 7.3) can provide insights into blender and material loading dynamics. The surface plots can highlight the need for assessing the appropriateness of the material loading process at the scale [2] as it can contribute toward variability in mixing and uniformity.

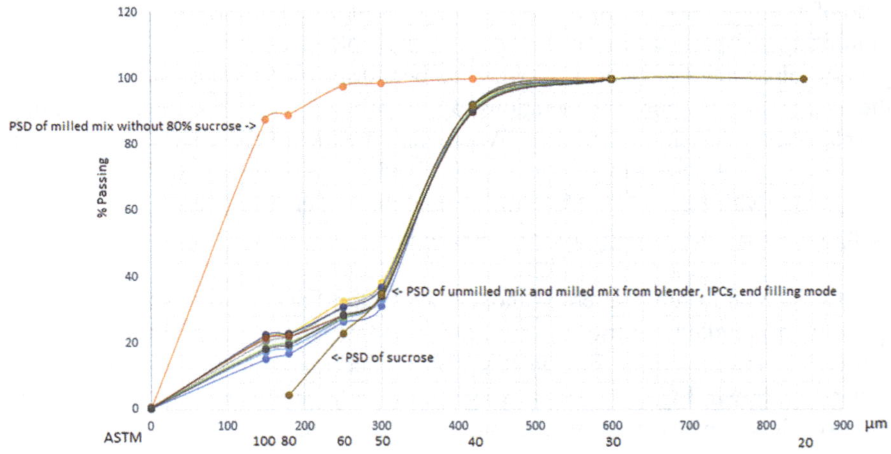

Fig. 7.2 PSD comparison example of the milled and unmilled mix

Fig. 7.3 A blender 3D surface plot. (Loaded blend before initial blending; after addition of remaining material; after final blend)

The blending mechanisms used for PfOS typically involve a V-blender, octagonal blender, or other types. In many legacy PfOS cases, tumbling dispersion octagonal blenders are used where the operations proceed with a low shear mixing [3]. In this case, the blender is rotated at a set speed and time. Tumbling dispersion octagonal blenders are suitable for gentle blending of dense granules/powders and abrasive materials that do not require breaking up mixing motion. In the dispersion blender, the particle motion is in a thin, cascading layer at the surface while the remainder of the material rotates with the vessel as a rigid body. As the blender size gets larger, the relative volume of particles in the cascading layer decreases and that may result in a decrease in mixing rate. The powder blend has a tendency for surface adherence of the layer in contact with the surface of the blender. For this reason, a powder flow disruptor such as a baffle or an adequate blender design (such as v-blender) is ideal [4]. Structured experiments have demonstrated that the addition of baffles has a positive impact on the mixing rate of powder blends. Disruption of regular flow patterns has shown beneficial effects on mixing rates. The permissible fill that does not create dead zones is also based on blender design/geometry and mixture character. The impact of blender design is more pronounced when the formulation is a fairly flowing dry powder for suspension as the case is in many PfOS (e.g., Hausner ratio of 1.20–1.27). The Stage 3 activities can include measurements, blend location uniformity sampling and testing, the powder rheology assessment, and discharge patterns observations (Table 7.3). They can be utilized to determine the adequacy of the equipment selection considerations (a current Stage 1 QbD prerequisite) during development of the legacy product. Where gaps are identified, design consideration such as Stage 1 DoE studies on a new blender equipment, a modified blender equipment (with effective baffles, knives, and intensifiers), or a lower scale blender equipment may be required for effective mixing and minimization of uniformity variability.

The material loading pattern is another key factor. For example, loading a higher quantity of unmilled sucrose on top of the milled material (which has API) and the subsequent impact during mixing may cause adhesion of material to the blender surface as shown (Fig. 7.4). The mixing motion in large blenders can create a rigid layer on blender walls. The adhered milled material (fines) may contain API that can then drop at the end of blending operation. The observation of blender adhered material and the corresponding location (e.g., T3 as seen in Fig. 7.4) results of the blend sampling may point toward drop of the adhered material at the end of the blending operation. The phenomenon may be confirmed based on consistent high results for the location across all Stage 3 batches. It can be further verified with the corresponding end of batch CU results. An optimal ratio of milled and unmilled bulking agent in the formulation and change in loading pattern (layering) can help in improving the observed phenomenon. Observing the blend for anomaly such as color, lump formation, or stickiness upon completion of blending operation is value adding.

Discharge Parameter Monitoring

The mix is typically unloaded into in-process containers (IPCs). Incremental dust if observed during material discharged may indicate segregation potential. The initial IPC discharge may allow for gravity material flow, while the final discharge may require manual interventions (Fig. 7.5). The dust observed signifies a higher amount of the glidant being discharged, and this happens when the discharged material is coming from a larger surface area within the blender. Protocol-driven Stage 3 material discharge observations can identify the continuous improvement needed to modify the blender design such that the dimensions/angles promote mass flow or to introduce alternate discharge mechanisms for legacy processes (Table 7.4). Such improvements may otherwise go unnoticed for a legacy process.

Table 7.3 An example of additional stage 3 blending parameter monitoring

Additional stage 3 Process monitoring plan	Additional stage 3 Sampling and testing plan	Rationale for 3 assessment
Unit operation: initial blending [octagonal blender]		
1. Verify initial blending parameters used 2. Measure distance between top of blender to surface of blend before start of blending 3. Measure distance between top of blender to surface of blend after initial blending	Not applicable	1. Determining potential source of variability 2. Enhancing legacy process knowledge
Unit operation: final blending [octagonal blender]		
1. Observe for any static in the polyethylene bag post charging 2. Verify blender functionality 3. Measure bin fill quantity 4. Ensure fixed blender parameters are used 5. Monitor order of addition 6. Ensure earthing to sampling tool 7. Prior to sampling, observe for any agglomeration 8. Observe blend uniformity sampling technique 9. Observe if valve is fully opened after unload 10. Ensure process continues to next stage without delay 11. Document distance between top of blender to surface of blend after final blending 12. Monitor transfer conditions and process	1. Three sets for assay from 10 different locations of the blender comprising of top (3), middle (3), and bottom (4) of the octagonal blender 2. Variance Component Analysis and Cpk Assessment to be performed on the results 3. Sample for water content (KF) 4. Sample for Bulk Density 5. Sample for Tapped Density 6. Sample for Compressibility Index & Hausner Ratio 7. Sample for PSD	1. Determining potential source of variability 2. Geometry of the blender is symmetric in design with distinct sections being the rectangular and hopper-shaped section 3. Inter/intra batch variability can be deduced from the current sampling plan T (3), M (3), B (4) 4. Enhancing legacy process knowledge

Filling Parameter Monitoring

The filling operation may need determination of adequacy to the legacy PfOS product. Determining the adequacy of a filler for filling a low active powder mix is more critical. For example, it is understood that auger fillers are more suitable for use with consistent particle distribution powder blends based on the design. The auger filler work mechanically crunching the particles with a possibility of over mixing. The selection of equipment in QbD approach is based on sound science and

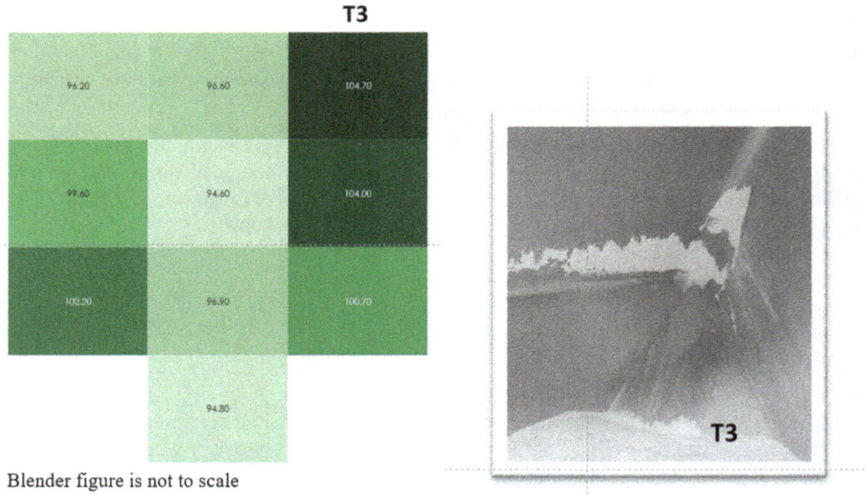

Fig. 7.4 Batch blend uniformity result

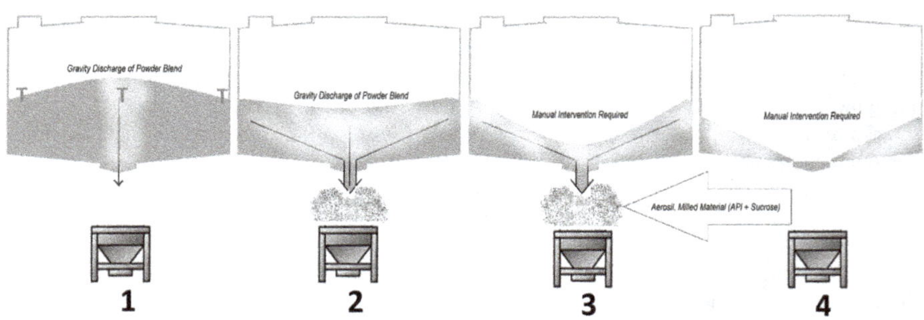

Fig. 7.5 IPC discharge pattern observed

manufacturers' overall level of product and process understanding to demonstrate control in achieving consistent and robust product. Being a legacy product, DoE studies are limited at Stage 1 to establish the design space for the product, wherein operating parameters are determined. For ungranulated powder blends, segregation can also happen during the discharge from IPC to hopper or during transfer by the auger mechanism. As larger the difference in particle size, the larger the potential exists. Segregation due to fluidization can occur when finer particles get aerated during the loading or unloading of the blend. This can be further pronounced with the auger filling mechanism. The heightened uniformity sampling and testing (with the application of ASTM E2709/E2810 and/or USP <905> AV) can gather supporting evidence of any filler variability (Table 7.5). The stratified sampling interval can be calculated based on process risk (i.e., more sampling at start and end) and on the actual weight of blend available for filling of each Stage 3 batch. The

Table 7.4 An example of additional stage 3 blend discharge monitoring

Additional stage 3 Process monitoring plan	Additional stage 3 Sampling and testing plan	Rationale for 3 assessment
Unit operation: blend discharge [IPCs]		
1. Document IPC fill sequence 2. Observe the unobstructed flow from the blender to IPCs 3. Observe for any air pockets during unload 4. Verify IPC fill quantities 5. Monitor IPC movements 6. Verify sampling tool earthing 7. Prior to sampling, observe for any agglomeration 8. Observe IPC sampling technique 9. Ensure process continues to next stage without delay 10. Monitor transfer conditions and process	1. Three sets for assay from 10 different locations of each of the 4 IPCs, comprising of top (3), middle (3), and bottom (4). 2. Variance component analysis and Cpk assessment to be performed on the results 3. Sample for water content (KF). 4. Sample for bulk density 5. Sample for tapped density 6. Sample for compressibility index & Hausner ratio 7. Sample for PSD	1. Determining potential source of variability 2. Generating additional data for the IPCs with 3 + 3 + 4 (T + M + B) sample locations 3. 3 + 3 + 4 (T + M + B) allows for determining inter/intra batch variability 4. Enhancing legacy process knowledge

results generated may help in determining if a stratified sampling and testing plan is required for commercial batches as a product control strategy.

The filler initial setting can be also verified during Stage 3. During initial setup of the powder filling machine, the fill range can be set to be within 0.5% (or other, as applicable) of the target fill weight as a starting point for machine setup. The IPCs, if multiple, may be loaded on to the powder filling machine sequentially for traceability and monitor if the IPCs are replaced immediately one after the other. Discharge to an empty hopper may increase the potential for blend segregation. The bottle reject quantities are to be observed throughout the Stage 3 batches.

Stage 3 Data Analysis

Nonparametric correlation measures such as Spearman's p, Kendall's tau, and Hoeffding's D can be used to compare the powder mix's physical attributes to uniformity outcomes. In one case, the analysis determined the correlation between

Table 7.5 An example of additional stage 3 filling operations monitoring

Additional stage 3 Process monitoring plan	Additional stage 3 Sampling and testing plan	Rationale for 3 assessment
Unit operation: powder filling [vacuum filler]		
1. Verify filler functionality 2. Document IPC use sequence 3. Record any additional IPC movement 4. Use defined pre-setup parameters 5. Document vibrator speed, dosing, and dosing pressure 6. Monitor parameter changes made to get target fill weight 7. Observe sampling technique 8. Re-verification of check-weigher rejects 9. Verify vacuum and dust collector unit before and after 10. Verify sampling interval calculation 11. Observe if valve and vent opening is consistent for IPCs 12. Ensure continuous filling operations 13. Verify alarm logs	1. Risk-based stratified replicate sampling covering each of the six filler heads (10): Start, 2.5%, 5%, 25%, 50%, 75%, 92.5%, 95%, 97.5% and end fractions for testing by FP assay method 2. Risk-based stratified replicate sampling covering each of the six filler heads (10): Start, 2.5%, 5%, 25%, 50%, 75%, 92.5%, 95%, 97.5%, and end fractions for testing by BU method 3. Samples for bulk density 4. Samples for tapped density 5. Samples for compressibility index and Hausner ratio 6. Samples for PSD 7. Samples for water content 8. Start, middle, and end samples for viscosity 9. Dissolution profile testing	1. Determining potential sources of variability 2. Risk-based sampling plan with initial and end tailing samples 3. Representative samples across the batch 4. Sample size allows the application of ASTM criteria [1] 5. To determine intra and interlocation variability

the % of fines and IPC BU results; when more fines were discharged, IPC BU results were high. This was consistent with the process observation of different discharge patterns for IPCs. In another assessment, it showed that there was a significant positive correlation confirming that manual discharge valve operations (# of opening and closing of the discharge valve) affect the blend homogeneity. A higher number of discharge valve operations corresponded to better uniformity results. Such monitoring data and observations clearly suggest the need for developing a controlled discharge valve mechanism for powder blend discharge operations. Segregation can occur when the blend is discharged from the blender into the downstream IPC. As the powder flows from the blender into a receiving container, it typically forms a heap. Particles rolling down the heap's slope often exhibit sifting segregation, in which smaller particles fall vertically into the interstices between larger particles and concentrate near the heap's center. Larger particles cannot fall straight down to the same extent and instead roll down, thus accumulating around the heap's perimeter. When the powder heap in the container is emptied into the next process unit (i.e., the hopper), the blend's composition and particle size distribution may fluctuate. The tendency may be minimized by changes such as reducing the number of IPC containers (e.g., blender scale down or use by use of a larger IPC). A larger number of operators may be required for the blend discharge operation, and intermittent interruption may be needed if excessive dust is generated during the drop. Adequate containment and a controlled powder blend discharge mechanism can be employed to avoid sudden drop and disruption of the blend are required. Such process improvements can be easily observed, and change initiated is based on the Stage 3 monitoring documentation.

The heightened sample size of IPC samples may adequately determine the intra-inter location variability of the blend in its final container prior to filling. The uniformity results from the IPC container can also identify if pockets of active content exist in the blender post-blending operations. The IPC results can be verified if they correspond to the stratified CU fractions, an indication of control (minimal disruption) during discharge of the powder blend from IPC to the filler hopper during the powder filling operation. The results obtained from testing the duct-recovered material (across the Stage 3 batches) can provide insights if the recovered material consists of fines with API. An excessive recovered material may also point toward the occurrence of low-density fines causing the potential for segregation during the blend discharge or subsequent unit operations. Low-density fines tend to settle at the top. Containment of the dust and handling modifications may be required in such cases to minimize API loss (Fig. 7.6).

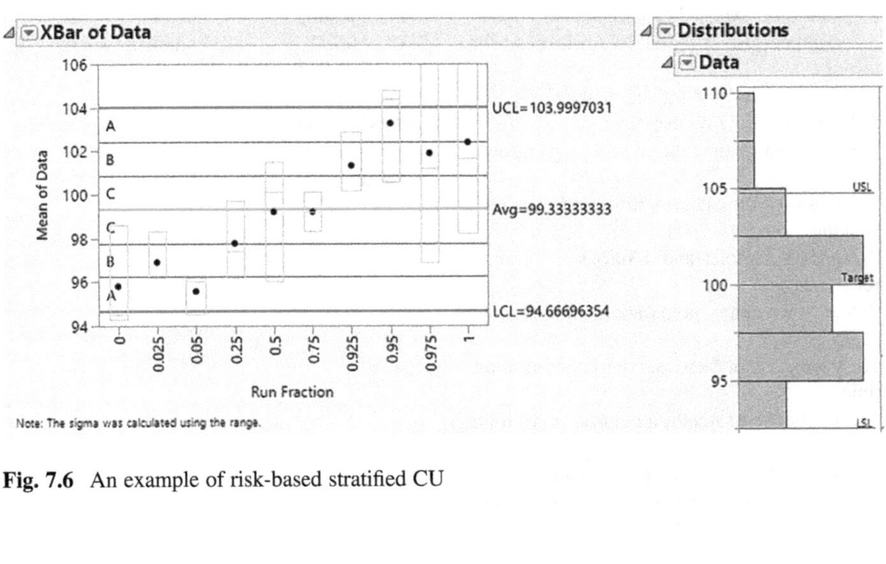

Fig. 7.6 An example of risk-based stratified CU

Determining Sources of Variability

The Stage 3 monitoring of unit operations may be executed in a continuous fashion, such that hold time variability is minimized while moving from one-unit operation to the other. With the addition of equipment functionality tests, the Stage 3 batches can provide supporting evidence on the consistency in functionality of the fluid bed dryer, sifter, mill, and the powder filler equipment for legacy products. Fixed processing equipment, earthing, and setup parameters may be used for all the previously determined number of Stage 3 batches. The Stage 3 monitoring can also document the training and operational consistency of manufacturing, packaging, sampling, and analytical personnel involved. The environmental conditions may be monitored through all Stage 3 batches with the heightened sampling and testing, such that any correlation may be determined (Table 7.6).

Analytical testing of the Stage 3 samples can be completed with monitoring (of weighing, preparation, instrument use, sample/standard handling, etc.) to identify sources of by analytical variability. Aspects such as refrigerated storage of prepared solution can be verified so that it minimize any degradation due to solution instability. The pattern of test results may be verified against individual analyst, HPLC, column, or mobile phase. Statistical comparison of, for example, CU SD (HPLC method) and the SD of dissolution (Q) (UV method) may provide insights

Table 7.6 An example of additional stage 3 analytical monitoring

Additional stage 3 Process monitoring plan	Additional stage 3 Sampling and testing plan	Rationale for 3 assessment
Analytical variability		
1. Verify use of new glassware 2. Track time exposed of prepared solution 3. Document use of the HPLC/instrument for testing 4. Document use the trained analyst 5. Document column use history 6. Document reconstitution and preparation time 7. Document the manner, time, and number of shaking 8. Verify if samples and dilutions are stored in refrigerator 9. Verify weighing, preparation, and instrument setup 10. Verify use of fixed standard conditioning time 11. Verify use of standard column conditioning time 12. Verify standard perpetration and expiry 13. Verify use of sonicator and if equal number of flasks are distributed in sonicator	1. BU HPLC method testing 2. CU HPLC method testing 3. Dissolution UV method testing 4. FP assay HPLC method testing	1. Determining potential source of variability 2. Enhancing legacy process knowledge

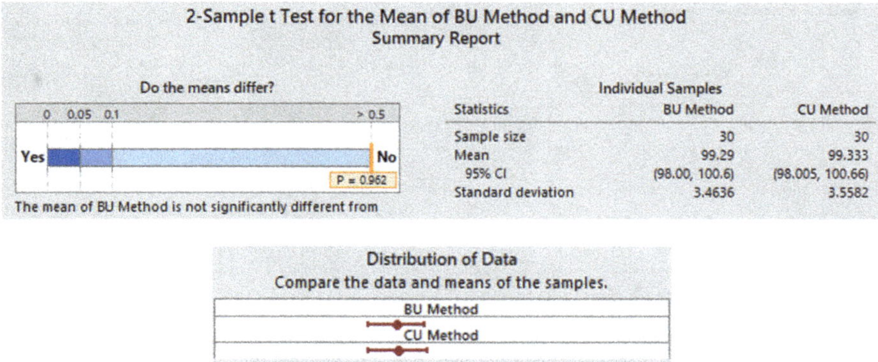

Fig. 7.7 Example of PfOS assay by BU and CU methods

previously unavailable. Another is the comparative assessment of assay results by the BU method (test method with direct weighing without reconstitution) and the CU method (test method with reconstitution step simulating actual patient use) if different (Fig. 7.7).

Variance Component Analysis of Uniformity Results

Blend uniformity results are analyzed to ensure homogeneity of the blend batch. Variance component analysis (VCA) is a method that assesses the amount of variation in a dependent variable that is associated with one or more random effects variables [5]. The variance components are typically calculated by three methods: restricted maximum likelihood (REML), Bayesian, and analysis of variance (ANOVA), which are dependent on the presence of an unbalanced or balanced sampling plan and negative estimated variance components. VCA results of the blend samples grouped by location area are used to determine the potential source of variability, within area and between areas (Fig. 7.8 and Table 7.7).

The above blender results indicated a similar degree of variance across all Stage 3 batches (% of total variance between batches is 5.4%). In total, 94.6% of the total blend variability across the Stage 3 batches was batch specific (within a batch). This is an indication that inadequate mixing is inherent to the blender and/or blending process; however, the pattern of occurrence is intrinsic to a batch. In the above case, first IPC container for all the Stage 3 batches had the lowest mean blend uniformity results while having the tightest SD. The data therefore is consistent with the potential for a swift gravity flow of the mix to the IPC, (coupled with observation of minimal dusting). This may highlight the inconsistency of blend discharge process into each IPC. Further in the case, high BU mean results were consistently observed in the final IPC across all Stag 3 batches. The observation of low (IPC # 1) and high (IPC # 4) uniformity results confirmed the existence of inadequate mixing

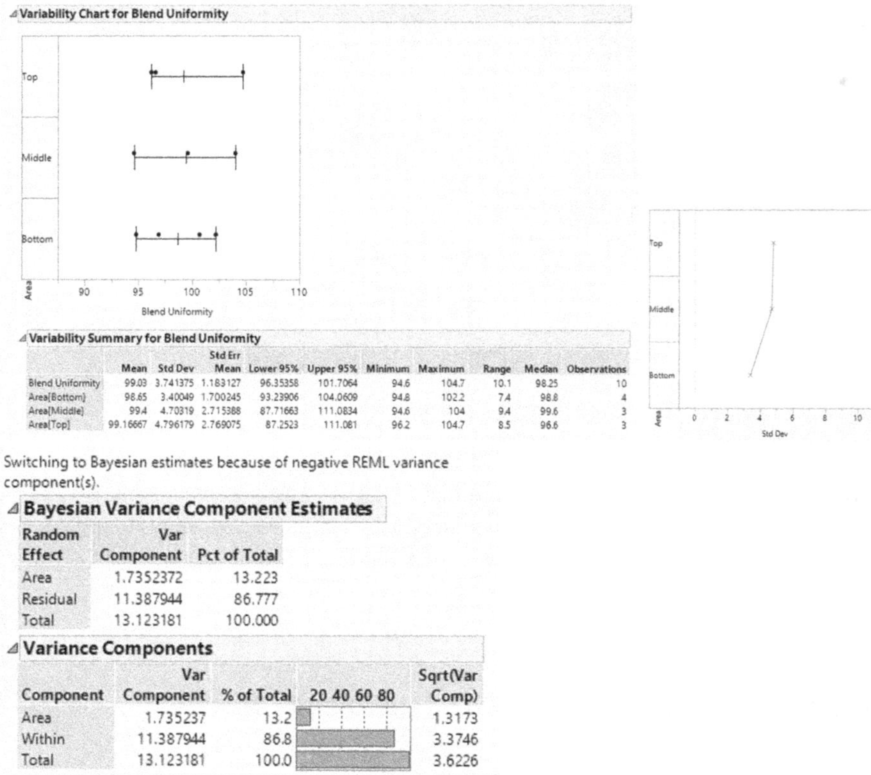

Fig. 7.8 A within and between blender area variability of a blend batch

Table 7.7 A variance components summary

Batch no.	Source of variability	Variance component	% of total variance	Square root of variance component
Pooled	Batch	0.53	5.4%	0.73
	Within	9.29	**94.6%**	3.05
	Total	9.82	N/A	3.13

of the blend and blend discharge variability. Since the IPC blend uniformity samples were more extensive than the blend uniformity samples from blender, the results adequately represent the entirety of the powder blend (Fig. 7.9). The Stage 3 heightened sampling and testing thus resulted in the immediate initiation of remedial measures (Stage 1) on the legacy product prior to observation of any product failure.

A replicate sample testing can negate the analytical method as a possible major contributing factor to variability [6]. More samples are taken deliberately per protocol from the inherently high-risk segments, i.e., beginning and end of the semi-continuous powder filling operation. Unlike batch operations, the variability of a semi-continuous processes is highly dependent on equipment capability and

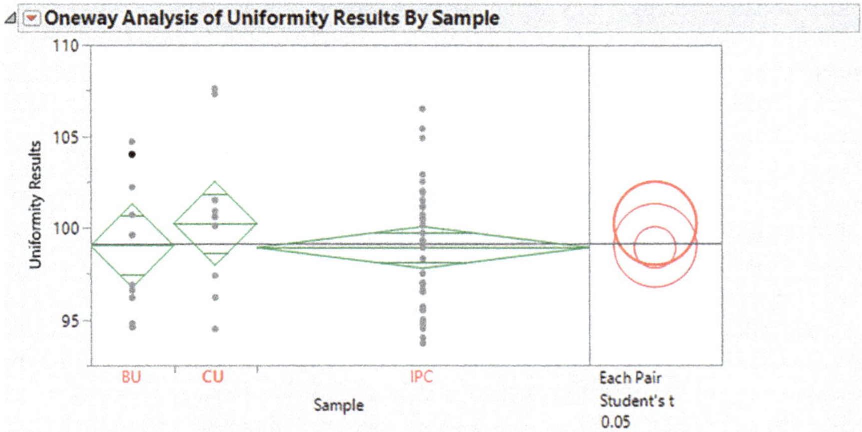

Oneway Anova

Summary of Fit

Analysis of Variance

Source	DF	Sum of Squares	Mean Square	F Ratio	Prob > F
Sample	2	14.06408	7.0320	0.5473	0.5815
Error	57	732.32575	12.8478		
C. Total	59	746.38983			

Means for Oneway Anova

Level	Number	Mean	Std Error	Lower 95%	Upper 95%
BU	10	99.030	1.1335	96.760	101.30
CU	10	100.230	1.1335	97.960	102.50
IPC	40	98.913	0.5667	97.778	100.05

Std Error uses a pooled estimate of error variance

Ordered Differences Report

Level	- Level	Difference	Std Err Dif	Lower CL	Upper CL	p-Value
CU	IPC	1.317500	1.267272	-1.22017	3.855168	0.3029
CU	BU	1.200000	1.602986	-2.00992	4.409924	0.4572
BU	IPC	0.117500	1.267272	-2.42017	2.655168	0.9265

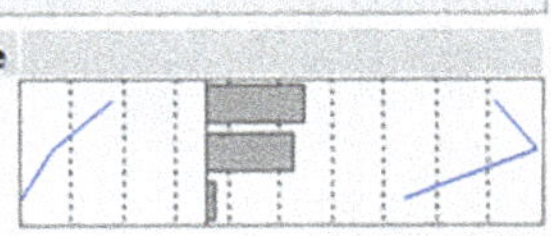

Fig. 7.9 Content uniformity, IPC and blender blend uniformity results of a batch

Table 7.8 Example of CU replicate results

St dev set 1	4.47 *(ASTM limit: 3.17)*
St dev set 2	4.15 *(ASTM limit: 2.90)*
St dev set 3	4.42 *(ASTM limit: 2.98)*

factors such as start/stop, beginning and end of operation, IPC changeover, etc. The risk-based stratified sampling plan for the Stage 3 heightened monitoring batches can being out any potential low/high content uniformity results at the beginning of the run and low/high results toward the end of the run. The results can be then related to the corresponding IPCs (Table 7.8).

The p-values for the replicate uniformity results can indicate the statistically significant difference in content uniformity data sets and IPC blend uniformity results at the 95% confidence level. The assessment can provide confirmation if there is any additional uniformity impact from the routine material transfer activities post blending, IPC to hopper discharge, powder filling process, or due to any localized hopper mixing. The assessment of BU, IPC BU, and CU results (p-values) for the Stage 3 batches can determine if the blender BU and IPC BU results can be a predictor for the CU results. Based on the assessment outcome, the use of CU as the sole uniformity control strategy may be justified.

The heightened sampling, testing, and observations during Stage 3 enable organizations to develop a greater understanding of the legacy product while inferring the sources of variability [7]. The assessment can be highly successful in detecting the presence and degree of variation and understanding the impact of variation on quality attributes. The insights from a heightened sampling and testing as part of Stage 3 monitoring of legacy products can be used to control the variation in a manner that is commensurate with the risk it represents to the process and product. The systematic observations, measurement, and analysis of data as part of such prognostic studies can result in developing a continuous improvement plan based on the fundamental QbD approach of applying sound pharmaceutical science principles supported by data. The heightened level of product and process knowledge gained provides support in making science- and risk-based decisions for the PfOS formulation and similar manufacturing processes and/or products. The study outcomes can be utilized for risk assessments [8], and Stage 1 design of experiments (focused product remediation). Organizations adopting such monitoring approaches for legacy products stand to benefit from agility [9, 10] in developing a fit-for-purpose continuous improvement/remediation program while fully realizing the potential of implementing process validation lifecycle Stage 3 approach as intended in the 2011 US FDA Process Validation Guidance for Industry.

References

1. ISPE Technical Team to assess Blend and Content Uniformity (BUCU): https://ispe.org/initiatives/blend-uniformity-content-uniformity/tools
2. Portillo PM, Ierapetritou M, Tomassone S, McDade C, Clancy D, Avontuur PC, Muzzio FJ (2008) Quality by design methodology for development and scale-up of batch mixing processes. J Pharm Innov 3:258–270
3. Bozzone S (2001) Solid oral dosage forms powder blending, Pfizer presentation, İKEV meeting
4. Muzzio FJ, Alexander AW (2005) Scale up of powder-blending operations, Pharmaceutical Technology s34
5. Sayeed-Desta N, Pazhayattil AB, Collins J, Doshi C (2018) A science and risk-based pragmatic methodology for blend and content uniformity assessment. AAPS PharmSciTech 19:1483–1492
6. Nunnally B (2009) Variance component analysis to determine sources of variation for vaccine drug product assays. J Valid Technol 15(3):78–88. https://www.proquest.com/openview/0f10310a4543d6cb1a767640564f8360/1?pq-origsite=gscholar&cbl=29232

7. Collins J, Sayeed-Desta N, Pazhayattil AB, Doshi C (2017) A novel metric for continuous improvement during stage three. Biopharm Int 30(6):32–35
8. Pazhayattil AB, Sayeed-Desta N, Chen S, Ingram M (2020) A semi-quantitative risk assessment methodology fit for bio-pharmaceutical lifecycle stages. PDA J Pharm Sci Technol 74(4): 423–434
9. Tomassini S, Pazhayattil AB, Ingram M (2020) A novel development and portfolio management strategy for generic small molecule drug products: Watergile ANDA. J Generic Med. https://doi.org/10.1177/1741134320918339
10. Pazhayattil AB, Konyu-Fogel G (2022) An empirical study to accelerate machine learning and artificial intelligence adoption in pharmaceutical manufacturing organizations. J Generic Med. https://doi.org/10.1177/17411343221151109

Appendices

Appendix I: Technology Transfer Documents (TTD) Checklist

Product Name:			
Strength:	**Dosage form:**		**Market:**
Sending Unit (SU):		**Receiving Unit (RU):**	
Shared by (SU QA): Sign/date		**Received By (RU QA): Sign/date**	

	Requirements	**Document**	**Shared (√/X/N)**	**Available (√/X/NA)**	**Remarks**	**Received by (RU QA) Sign/ Date**
1.	Test license/Mfg. license copy					
2.	Master formula card (MFC)					
3.	Material safety data sheets (MSDS)					
4.	Master batch manufacturing record (MBMR)					
5.	Operational packing material list (OPML)					
6.	Master batch packing record (MBPR)					
7.	Raw material specification (API & excipients) and standard test procedures.					
8.	In-process specification and standard test procedures.					

(continued)

A. B. Pazhayattil et al., *Technology Transfer*, AAPS Introductions in the Pharmaceutical Sciences 10, https://doi.org/10.1007/978-3-031-32192-4

9.	Finish product specification and standard test procedures					
10.	Packing material specifications and standard test procedures					
11.	Vendor details and vendor qualification reports. Approved supplier/s or manufacturer for raw and packing material					
12.	Stability study protocol					
13.	List of manufacturing/packing equipment and technical specification					
14.	Tool drawings (punches, blister layouts, etc.)					
15.	Stability study reports of exhibit/current commercial batches (accelerated/long term)					
16.	Latest annual product review (APR)					
17.	Cleaning validation protocol/report and test methods (if required)					
18.	Process validation/qualification protocol and reports					
19.	Analytical method validation data (if required)					
20.	Hold time data (if required)					
21.	Product dossier (if available)					
22.	Specific SOP related to process and cleaning					
23.	Filter validation data/documents (if applicable)					
24.	Others (if any specify)					

Evaluation:

…………………………………………………………………………………-
………………………………………………

CHECKED BY (RECEIVING UNIT) (SIGN/DATE)	APPROVED BY (RECEIVING UNIT) (SIGN/DATE)
MS&T	QUALITY ASSURANCE

Appendix II: Technology Transfer Protocol (TTP)

Table of Contents

Appendix III: Check-List for First Exhibit Batch Start-Up

Strength:	**Dosage Form:**	**Market:**
Customer:		
Sign/Date:		

Requirements	**Responsibility**	**Available**	**Sign/ Date**
		Yes/NA/ Remark	
Availability of approved raw materials and packing materials	GSO/MS&T		
Manufacturing license (if applicable)	QA		
Narcotics & psychotropic substances license and permission for the use of absolute alcohol (if applicable)	QA		
Vendor details and vendor approval document	QA		
Approved supplier/s or Manufacturer of raw & packing material	QA		

(continued)

Availability of raw material specifications (API & Excipients) and test procedures	QA		
Packing material specifications and test procedures	QA		
In-process specifications and test procedures	QA		
Finished product specifications and test procedures	QA		
Cleaning validation and validated cleaning test methods	QA		
Any specific product handling instruction for environmental controls and safety (respirator/nose mask, etc.)	EHS		
Analytical method Validation/Analytical method Transfer (API & FP)	QC/QA		
Hold time study protocol	QC/QA		
Technical agreement (if applicable)	QA		
Master formula card (MFC)	MS&T/QA		
Scale up data	MS&T		
Equipment requirements and, if different than scale up equipment, equipment qualification, and suitability assessment	MS&T/RA		
Product stability protocol	MS&T/QA		
Process qualification/validation protocol, including CMA, critical quality attributes, and CPP (as applicable)	MS&T/QA		
Operational packing material list	PKG/QA		
Master BMR	PDL/QA		
Master BPR	PKG/QA		
Equipment change parts availability 1. Tooling: 2. Blister layouts and change parts	PR/PKG		
Pack details	PKG		
Approved artwork of packing material	PKG		
Filter validation data/documents (if applicable)	MS&T/QA		
Regulatory impact assessment	RA		
Other (if any specify)			

REVIEWED BY								APPROVED
MS&T	PKG.	PR	GSCO	WH	EHS	QC	RA	QA

Appendix IV: Case Study of Roller Compaction Process from Development to Scale-Up

Target Product Profile The pharmaceutical target profile for any drug product should be a safe, efficacious convenient dosage form that should facilitate patient compliance. Here the tablet dosage form is discussed with the roller compaction process. The tablet should be of an appropriate size, with a single tablet per dose. The manufacturing process for the tablet should be robust and reproducible and result in a product that meets the appropriate drug product critical quality attributes. A target product profile mentioned in Table 1 details the critical quality attributes, which are used to define the satisfactory quality parameters identified.

This pertains to developing a drug product (DP) with a roller compaction process. The composition risk concluded from the development of the drug product is mentioned in Table 2.

Based on the development study, the concluded component levels and attributes are listed in Table 3

Table 1 Target product profile

Quality attribute	Target	Criticality
Dosage form	Tablet, maximum weight 200mg	Not applicable
Potency	30 mg	Not applicable
Pharmacokinetics	Immediate release enabling Tmax in 2 hours or less	Related to dissolution
Appearance	Tablet conforming to description shape and size	Critical
Identity	Positive for drug substance	Critical
Assay	95–105%	Critical
Impurities	Impurity A: NMT 0.5%, Other impurities: NMT 0.2%, Total: NMT 1%	Critical
Water	NMT 1%	Not critical – API not sensitive to hydrolysis
Blend Uniformity	10 location SD should NMT 3	Critical
Content Uniformity	Meets ASTM 2810 criteria	Critical
Resistance to Crushing (Hardness)	50–120 N	Not critical
Friability	NMT 1.0%	Not critical
Dissolution	Consistent with immediate release, e.g., NLT 80% (Q) at 30mins	Critical
Disintegration	NMT 15mins	Not critical, a precursor to dissolution
Microbiology	If testing is required, meets USP criteria	Critical only if a drug product supports microbial growth

Table 2 DP formula composition risk identified after development at lab scale

	Formulation Composition Attributes						
DP CQA	API Particle Size	API Level	Lactose Level	Disintegrant Level	MCC Particle Size	Glidant Level	Magnesium Stearate Level
Appearance	Low	Low	Low	Low	Low	Low	High
Identity	Low	Low	Low	Low	Low	Low	Low
Assay	Low	Low	Low	Low	Low	Low	Low
Impurity	Low	Low	Low	Low	Low	Low	Low
Blend Uniformity	High	High	Low	Low	High	Low	Low
Content Uniformity	High	High	Low	Low	High	Low	Low
Dissolution	High	High	Low	High	Low	Low	High

Table 3 Summary of outcome of formulation components study

Drug substance particle size	D90 35–45 microns
API concentration	14.28% w/w
Croscarmellose level (Disintegrant)	3–4% w/w
Magnesium stearate level	1–2% w/w (Intragranular) 0.25% w/w (Extra granular)
Microcrystalline cellulose level	40% w/w
Lactose monohydrate level	39.00–40.75%
Talc level (Glidant)	5%

Summary of the Selected Process After Development

In this case study, based on the physicochemical properties of the API, roller compaction was selected as the most appropriate manufacturing process. The API is sensitive to heat which would preclude wet granulation due to chemical instability during the drying process. In addition, the API physical properties (flow) precluded direct compression at the concentrations required. Tablet coating was also precluded due to chemical instability during drying. A flow diagram of the manufacturing process for the drug product is provided in Fig. 1. Microcrystalline cellulose, lactose

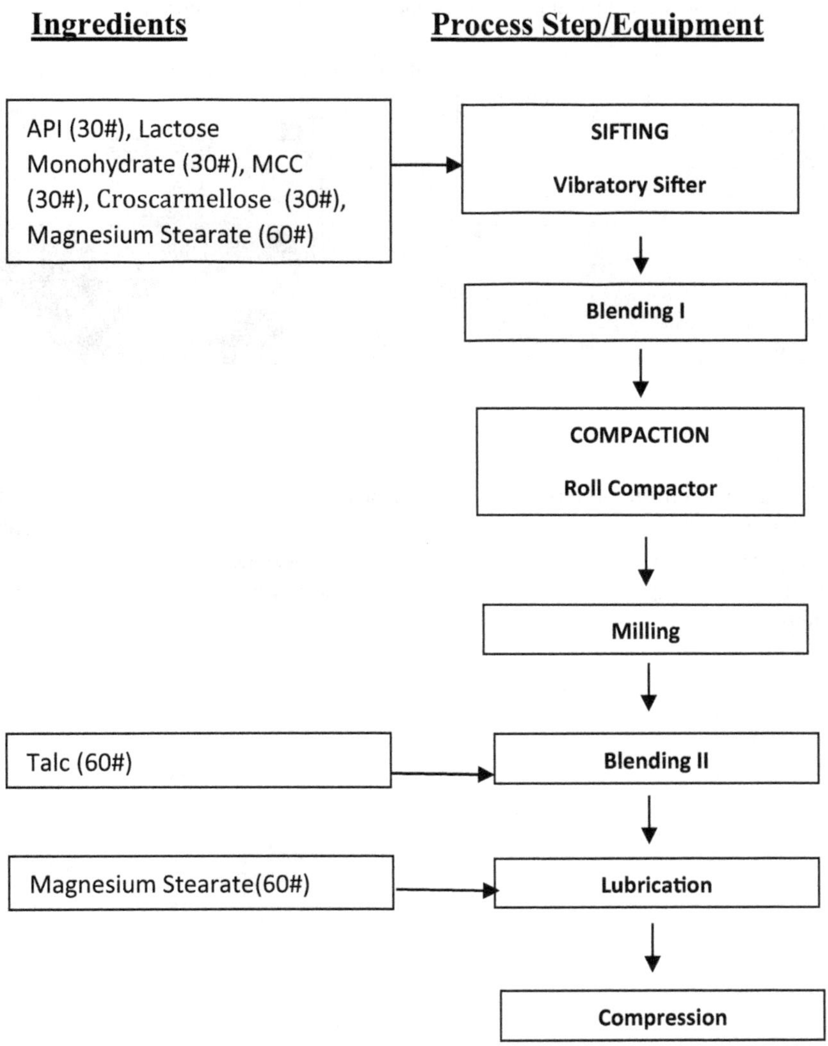

Fig. 1 Process flow diagram

monohydrate, croscarmellose sodium, and magnesium stearate are separately weighed and screened and then blended with sifted API. The blend is then roller-compacted to produce a ribbon, milled to give active granules. Extragranular ingredients (magnesium stearate and talc) are separately weighed and screened and then blended with the granules. The blend is then compressed into tablets.

Table 4 Risk assessment to identify variables potentially impacting product quality

DP CQA	Blending I	Roller Compaction	Milling	Blending II	Lubrication	Compression
Appearance	Low	Low	Low	Low	Low	High
Identity	Low	Low	Low	Low	Low	Low
Assay	Low	Low	Low	Low	Low	High
Impurity	Low	Low	Low	Low	Low	Low
Blend Uniformity	High	High	High	High	High	High
Content Uniformity	High	High	High	High	High	High
Dissolution	Low	High	High	Low	High	High

Based on scientific understanding and prior knowledge, a risk assessment of the potential impact of the unit operations on the drug product CQAs was identified after development. Table 4 shows the result of the risk assessment and identifies the unit operations that require further investigation to determine the appropriate control strategy.

Index

A. B. Pazhayattil et al., *Technology Transfer*, AAPS Introductions in the Pharmaceutical Sciences 10, https://doi.org/10.1007/978-3-031-32192-4

www.ingramcontent.com/pod-product-compliance
Lightning Source LLC
Chambersburg PA
CBHW070050060125
19957CB00003B/99

9 783031 321917